UNITEXT for Physics

UNITEXT for Physics series publishes textbooks in physics and astronomy, characterized by a didactic style and comprehensiveness. The books are addressed to upper-undergraduate and graduate students, but also to scientists and researchers as important resources for their education, knowledge, and teaching.

Roberto A. Capuzzo Dolcetta

Classical Newtonian Gravity and Potential Theory

A Comprehensive Introduction, with Examples and Exercises

Second Edition

Springer

Roberto A. Capuzzo Dolcetta
Dipartimento di Fisica
Sapienza Università di Roma
ROMA, Roma, Italy

ISSN 2198-7882 ISSN 2198-7890 (electronic)
UNITEXT for Physics
ISBN 978-3-032-13035-8 ISBN 978-3-032-13036-5 (eBook)
https://doi.org/10.1007/978-3-032-13036-5

This Springer imprint is published by the registered company Springer Nature Switzerland AG
The registered company address is: Gewerbestrasse 11, 6330 Cham, Switzerland

To Alvio, with love and pain, and to Daniela and Flavio, with love and gratitude.

Preface

This book is a revised and expanded edition of my earlier volume with a similar title, published in 2019 in the Unitext for Physics series. In particular, Chap. 2 has been substantially revised and expanded, while Chaps. 4 and 6 are entirely new.

The purpose of this book is to collect and clarify many topics which, though basic, are not easily found together in a single textbook and are often less trivial than they appear.

This book aims to provide an accessible yet comprehensive introduction to classical Newtonian theory of gravitation and of potential theory, as fundamental pillars of physics that are of particular relevance to astrophysics.

It is known, indeed, that among the four fundamental forces of Physics, gravity has the unique aspect of being an unscreened force, which permeates the whole universe at all scales. On the other hand, in addition to the above property (which is often referred to as *infrared* divergence), Newtonian interaction among individual point masses is characterized, also, by the *ultraviolet* divergence: the force exerted by one point mass on another is unlimited when the two bodies approach too close to each other. These two characteristics of Newtonian gravitation are not real issues for distributions of matter that are limited in size and more regular (i.e., extended objects rather than point masses).

Another important conceptual limitation of purely Newtonian gravitational interaction is the intrinsic assumption that it produces a force field whose intensity and direction at every point in space instantly respond to changes in the positions of all massive bodies. This corresponds to an *infinite* interaction speed, which is unlikely and was strongly criticized by Einstein.

In spite of these practical and conceptual limitations, Newtonian gravitation is an exceptional tool for the understanding of the mechanics on the Earth and in the sky over an enormous range of space and time scales and over a wide range of velocities. In other words, whenever both the approximation of a *weak* gravitational field and that of *low* velocities (with respect to the speed of light in vacuum, c) of the objects under study are valid, General Relativity (or similar theories) and Special Relativity are irrelevant to a precise determination of the objects' motion. For example, on the Earth, the consideration of General Relativity effects has a role just when an extremely high accuracy of position determination is required (GPS systems). On the other hand, Special Relativity has no practical relevance for low-speed mechanics on macroscopic scales; its effects appear only when velocities are a substantial fraction of c. Things are different in an astrophysical context, due to the presence of compact objects like massive black holes generating gravitational fields so intense as to require general and special relativity theories for a proper dynamical study of their environment.

It would be superfluous to recall the exceptional successes of classical Newtonian gravitation theory as a tool to launch and control with enormous precision interplanetary probes, at least as long as the probe remains sufficiently far from the Sun. Let us remember that even for the closest planet to the Sun, Mercury, the deviation from the purely classical determination of its perihelion precession is just $\sim 43''$ per tropical century. This deviation, accounted for by General Relativity, amounts to only 0.8% of the total measured longitude precession which is $5600''$ per tropical century.

So the classical celestial mechanics (few-body problem) up to stellar dynamics (large N-body problem) can be treated with high precision in the frame of classical Newtonian dynamics, at least when the distances between the bodies are large compared with the Schwarzschild radii of the moving objects.

With this book, I hope to have provided a mathematically reliable yet simple treatment of the main issues in classical Newtonian gravitational theory for particle systems and regular distributions of matter. To this end, I have included a series of exercises, most of them with complete solutions.

This book is primarily intended for undergraduate and graduate students in physics, mathematics, and astronomy or astrophysics. However, it should also appeal to any reader with a college-level background in mathematics and physics.

Roma, Italy
May 2026

Roberto A. Capuzzo Dolcetta

Acknowledgements I wrote some parts of this book while I was a guest at the Institut für Theoretische Physik der Universität Heidelberg (Germany), the institution whose hospitality I warmly acknowledge.

Competing Interests The author has no competing interests to declare that are relevant to the content of this manuscript.

Contents

Symbols

List of symbols used in this book.

$\mathrm{M}_\odot$	Sun mass ($1\ \mathrm{M}_\odot = 1.9891 \times 10^{30}$ kg).
$\mathrm{R}_\odot$	Sun radius ($1\ \mathrm{R}_\odot = 6.9570 \times 10^{5}$ km).
G	Newtonian constant of gravitation ($G = 6.67408 \times 10^{-11}$, in MKS units).
AU	Astronomical Unit, i.e. the average Sun-Earth distance: 1 AU= 1.496×10^{8} km.
e	Euler's number, base of natural logarithms, e $\simeq$ 2.71828.
$\\|\mathbf{v}\\|$	With this symbol we intend the Euclidean norm of a vector $\mathbf{v} = \sum_{i=1}^{n} v_i \mathbf{e}_i$, in $\mathbb{R}^n$. This means $\\|\mathbf{v}\\| = \left(\sum_{i=1}^{n} v_i^2 \right)^{1/2}$. Alternatively, we use also v to indicate the norm of $\mathbf{v}$.
$(r,\ \theta,\ \varphi)$	Spherical polar coordinates of a point in $\mathbb{R}^3$.
$(R,\ \theta,\ z)$	Cylindrical coordinates of a point in $\mathbb{R}^3$.
$\mathbb{R}$, $\mathbb{Z}$, $\mathbb{Z}^{\pm}$	With these symbols we refer, respectively, to the set of real, integer, and positive (+) or negative (−) integer numbers.
$\mathbf{e}_i$	This symbol refers to the unit vector in the ith coordinate direction in $\mathbb{R}^n$.
$\mathbf{e}_v$	Unit vector in the direction of generic vector $\mathbf{v}$, $\mathbf{e}_v \equiv \mathbf{v}/v$.
$\mathbf{r}$	Represents the position vector in $\mathbb{R}^n$, i.e. $\mathbf{r} = \sum_{i=1}^{n} x_i \mathbf{e}_i$.
$x_i v_i$	Einstein's summation convention, that implies sum over repeated indices, i.e., in $\mathbb{R}^n$, $x_i v_i \equiv \sum_{i=1}^{n} x_i v_i$.
$\mathbf{n}$	This symbol indicates the outward unit vector in a point of a surface.
$\cdot$, $\wedge$	These symbols represent, respectively, scalar ($\cdot$) and vector ($\wedge$) products between vectors.
$\in$	Symbol for "belongs to".
$\forall$	Symbol for "for all".
$\exists$	Symbol for "exists".

$\parallel$	Indicates geometrical parallelism (do not confuse with the double $\|$ for norm).
$\widehat{\mathbf{u}\,\mathbf{v}}$	Angle, in the interval $[0, \pi]$, between vectors $\mathbf{u}$ and $\mathbf{v}$, also expressed as $\cos\widehat{\mathbf{u}\,\mathbf{v}} = (\mathbf{u}\cdot\mathbf{v})/(\Vert\mathbf{u}\Vert\,\Vert\mathbf{v}\Vert)$.
∇	In $\mathbb{R}^n$, this symbol, named *nabla*, represents the formal vector operator $\nabla \equiv \sum_{i=1}^{n} \mathbf{e}_i \frac{\partial}{\partial x_i}$.
$\setminus$	Symbol for difference between sets.
$\det(\underline{\mathbf{A}})$	Determinant of matrix $\underline{\mathbf{A}}$, also written as $\vert\underline{\mathbf{A}}\vert$.
$S_R(\mathbf{r})$	With this symbol we indicate, in $\mathbb{R}^n$, the spherical hypersurface of radius R and centre at $\mathbf{r}$.
$E, \mathbf{F}, \mathbf{Q}, \mathbf{L}$	These symbols represent, respectively, the mechanical energy, force, linear momentum, and angular momentum, all per unit mass.
$[F(x)]_a^b$	Indicates $F(b) - F(a)$, where $F(x)$ is a function of one variable.
$\sum_{\substack{(i,j)=1 \\ i\neq j}}^{N} X_{ij}$	Sum over the 2 indices of the $N \times N$ array X_{ij}, excluding $i = j$.
$[x]$	Integer part of x, i.e. the greatest integer less than or equal to x.
$a^{\pm}$	The notations a^+ and a^- indicate that a limit approaches the value a from above or below, respectively.
$n!$	Is the factorial of n: $n! = \prod_{k=0}^{n-1}(n-k)$.
$n!!$	Is the double factorial of an integer $n \geq 0$, i.e. the product of all the positive integers up to n that have the same parity (odd or even) as n: $n!! = \prod_{k=0}^{[n/2]-1} (n-2k)$.
$\binom{n}{k}$	For integers n and k, $0 \leq k \leq n$, it gives the binomial coefficient $\frac{n!}{k!(n-k)!}$.
$\left(\binom{\alpha}{k}\right)$	For real α and integer k, $0 < k \leq \alpha$, it gives the *generalized* binomial coefficient: $\left(\binom{\alpha}{k}\right) = \frac{\alpha(\alpha-1)\cdots(\alpha-k+1)}{k!}$. This reduces to $\binom{\alpha}{k}$ when α is a positive integer.
$\ln x$	Represents the natural logarithm (in base e) of $x > 0$.
$\log x$	Represents the decimal logarithm (in base 10) of $x > 0$.
$\mathcal{C}^k$	Represents the space of functions continuous up to the kth derivative.
$L^2([a, b])$	Represents the space of square-integrable functions in the interval $[a, b]$.
∞^k	Indicates that there are k independent arbitrary choices for a parameter.
$\Longrightarrow$	Logical implication: $a \Longrightarrow b$ means *a implies b*, i.e. that if proposition a is true, then proposition b is also true.

$\Longleftarrow$ Converse logical implication: $a \Longleftarrow b$ means b *implies* a, i.e. that if proposition b is true, then proposition a is also true.

$\Longleftrightarrow$ Double logical implication, that means identity between propositions a and b.

Introduction

This book is the second edition, corrected and expanded, of the volume originally published in 2019. This second edition includes a substantially revised Chap. 2 and two entirely new Chaps. (4 and 6).

My goal with this book is to provide an introduction to classical Newtonian gravitation and potential theory, fundamental components of physics essential for understanding classical mechanics and particularly relevant to astrophysics.

Among the four fundamental forces of nature, gravity has the unique aspect of being an unscreened force that permeates the entire universe. Although Einstein's general relativity theory provides a broader framework for gravity, Newtonian theory remains much simpler and sufficiently accurate for most practical and scientific applications, provided the gravitational field is weak.

To achieve its goal, the book is organized as follows.

Chapter 1 recalls some essential elements of vector calculus, especially to introduce the formalism used in the following chapters.

Chapter 2 presents and discusses the classical Newtonian theory of gravitation for one and for an arbitrary number N of point masses. The chapter contains extensive considerations on the classical Newtonian N-body problem, dealing with the virial theorem, the concepts of relaxation time and of *collisionality*, and their consequences on the overall dynamics. The theory for point masses is naturally extended to the continuous case in the second part of the chapter.

In the third chapter, the paradigmatic case of spherical symmetry in the mass density distribution (central force) is treated, with the introduction of a useful method of qualitative analysis of motion.

Chapter 4 presents some astrophysical applications of Newtonian gravitation, beginning with the definitions and properties of self-gravitating systems. The study of the equilibrium and non-equilibrium behaviour of self-gravitating spherical fluids follows. In this context Jeans's theory of gravitational instability is introduced, as a first step towards the understanding of star formation in galaxies.

In Chap. 5 the general case of a non-spherical mass density distribution is discussed. In this chapter, classical potential theory is presented, with elements of harmonic theory, essentially to understand the series expansion of the potential discussed in the second part of the chapter. The final part of this chapter deals with the specific case of the motion of a satellite around the Earth.

The sixth and final chapter examines some of the weak points and issues involving Newtonian gravitation and their solution within the framework of special and general relativity theories.

Examples and exercises are presented throughout the book to clarify aspects of the theory.

Chapter 1
Elements of Vector Analysis and Calculus

This chapter is devoted to elements of vector analysis and calculus, which provide the fundamental knowledge and tools needed to deal with the material of the following chapters.

We begin with the theory of vector functions of real variables and then introduce the formal *nabla* operator, ∇, together with its properties and applications in various coordinate systems. The chapter continues with vector field theory and its fundamental differential theorems, including the divergence theorem, Stokes' theorem, Gauss' theorem, and the continuity equation.

1.1 Vector Functions of Real Variables

Let $\mathbf{v} = \mathbf{v}(t)$ be a real function of one real variable.

If $\mathbf{v} \in \mathbb{R}^3$, we may write

$$\mathbf{v}(t) = v_x(t)\mathbf{i} + v_y(t)\mathbf{j} + v_z(t)\mathbf{k}, \tag{1.1}$$

where $\mathbf{i}$, $\mathbf{j}$, $\mathbf{k}$ are the unit vectors in the directions of the axes of a Cartesian orthogonal coordinate system, such that

$$\mathbf{i} \cdot \mathbf{i} = \mathbf{j} \cdot \mathbf{j} = \mathbf{k} \cdot \mathbf{k} = 1, \tag{1.2}$$

$$\mathbf{i} \cdot \mathbf{j} = \mathbf{i} \cdot \mathbf{k} = \mathbf{j} \cdot \mathbf{k} = 0. \tag{1.3}$$

Example 1.1.1 The circle of radius R is represented with the function

$$\mathbf{r}(\theta) = R\cos\theta\,\mathbf{i} + R\sin\theta\,\mathbf{j},$$

with $0 \leq \theta < 2\pi$.

R. A. Capuzzo Dolcetta, *Classical Newtonian Gravity and Potential Theory*, UNITEXT for Physics, https://doi.org/10.1007/978-3-032-13036-5_1

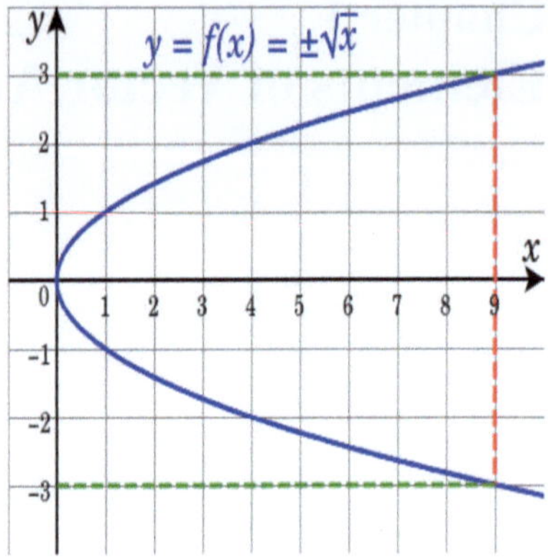

Fig. 1.1 A double-valued function plot

Example 1.1.2 The Cartesian and parametric representations of a parabola are

$$y = ax^2, \qquad \begin{cases} x = t, \\ y = at^2. \end{cases}$$

In vector formalism, it corresponds to a mapping $\mathbf{r} : \mathbb{R} \longrightarrow \mathbb{R}^2$ expressed as

$$\mathbf{r} = t\ \mathbf{i} + at^2\ \mathbf{j}.$$

Note that hereafter we will deal only with single-valued vector functions, excluding multi-valued functions, which do not have an inverse function (Fig. 1.1).

1.2 Limits of Vector Functions

Given the vector function $\mathbf{v} = \mathbf{v}(t)$, if

$$\forall \epsilon > 0, \quad \exists \delta_\epsilon > 0 \quad \text{such that for } |t - t_0| < \delta_\epsilon, \tag{1.4}$$

then

$$\|\mathbf{v}(t) - \mathbf{V}\| < \epsilon, \tag{1.5}$$

we say that

$$\lim_{t \to t_0} \mathbf{v}(t) = \mathbf{V}. \tag{1.6}$$

Note that, if $\mathbf{v}(t) = v_x(t)\,\mathbf{i} + v_y(t)\,\mathbf{j} + v_z(t)\,\mathbf{k}$, then

$$\lim_{t \to t_0} \mathbf{v} = \mathbf{V} \Longleftrightarrow \begin{cases} \lim_{t \to t_0} v_x = V_x, \\ \lim_{t \to t_0} v_y = V_y, \\ \lim_{t \to t_0} v_z = V_z. \end{cases} \tag{1.7}$$

A vector function $\mathbf{v} = \mathbf{v}(t)$ is said to be continuous in t_0 if

$$\lim_{t \to t_0} \mathbf{v}(t) = \mathbf{v}(t_0). \tag{1.8}$$

Clearly, the function $\mathbf{v}(t)$ is continuous if and only if all its components, $v_x(t)$, $v_y(t)$, $v_z(t)$, are continuous. The usual properties of scalar continuous functions still hold for vector functions:

- If $\mathbf{v}$ is continuous then $\alpha\mathbf{v}$, where $\alpha \in \mathbb{R}$, is also continuous.
- If $\mathbf{v}$ and $\mathbf{w}$ are continuous then $\mathbf{v} \cdot \mathbf{w}$ is also continuous.
- If $\mathbf{v}$ and $\mathbf{w}$ are continuous then $\mathbf{v} \wedge \mathbf{w}$ is also continuous.
- If $\mathbf{v}$ and $\mathbf{w}$ are continuous then $\mathbf{v} \pm \mathbf{w}$ is also continuous.

Moreover, some additional properties for limits of vector functions hold.
Let $\mathbf{v}(t)$ and $\mathbf{w}(t)$ two vector functions such that

$$\lim_{t \to t_0} \mathbf{v}(t) = \mathbf{V}, \qquad \lim_{t \to t_0} \mathbf{w}(t) = \mathbf{W}, \tag{1.9}$$

then

$$\begin{aligned} &\bullet \lim_{t \to t_0} \alpha\mathbf{v} = \alpha\mathbf{V} \quad (\alpha \in \mathbb{R}), \\ &\bullet \lim_{t \to t_0} \|\mathbf{v}\| = \|\mathbf{V}\|, \\ &\bullet \lim_{t \to t_0} \mathbf{v} \cdot \mathbf{w} = \mathbf{V} \cdot \mathbf{W}, \\ &\bullet \lim_{t \to t_0} \mathbf{v} \wedge \mathbf{w} = \mathbf{V} \wedge \mathbf{W}. \end{aligned} \tag{1.10}$$

1.3 Derivatives of Vector Functions

Let $\mathbf{v} = \mathbf{v}(t)$ be a vector function of a real variable t defined in an interval I. For any arbitrary point $t \neq t_0$ in I let us consider the difference quotient

$$\frac{\mathbf{v}(t) - \mathbf{v}(t_0)}{t - t_0}. \tag{1.11}$$

If the difference quotient (1.11) has a finite limit

$$\left(\frac{\mathrm{d}\mathbf{v}}{\mathrm{d}t}\right)_{t_0} \equiv \dot{\mathbf{v}}_0 = \lim_{t\to t_0} \frac{\mathbf{v}(t) - \mathbf{v}(t_0)}{t - t_0}. \tag{1.12}$$

(the dot over the vector is hereafter indicating derivative with respect to t) we say that the function $\mathbf{v} = \mathbf{v}(t)$ is differentiable in the point t_0 of the domain I, and such limiting value is called derivative of $\mathbf{v}(t)$ in t_0.

The following relations hold:

$$\begin{aligned}
&(i)\ \frac{\mathrm{d}}{\mathrm{d}t}(\alpha\mathbf{v}) = \alpha\frac{\mathrm{d}}{\mathrm{d}t}\mathbf{v}, \quad \forall\alpha \in \mathbb{R},\\
&(ii)\ \frac{\mathrm{d}}{\mathrm{d}t}(\mathbf{v} + \mathbf{w}) = \dot{\mathbf{v}} + \dot{\mathbf{w}},\\
&(iii)\ \frac{\mathrm{d}}{\mathrm{d}t}(\mathbf{v} \cdot \mathbf{w}) = \dot{\mathbf{v}} \cdot \mathbf{w} + \mathbf{v} \cdot \dot{\mathbf{w}},\\
&(iv)\ \frac{\mathrm{d}}{\mathrm{d}t}(\mathbf{v} \wedge \mathbf{w}) = \dot{\mathbf{v}} \wedge \mathbf{w} + \mathbf{v} \wedge \dot{\mathbf{w}}.
\end{aligned} \tag{1.13}$$

Theorem 1.1 *If the vector function $\mathbf{v}(t)$ is differentiable in t_0, it is continuous in t_0.*

Proof Let $\mathbf{v}(t)$ differentiable in the point $t = t_0$. It implies that the limit

$$\lim_{t\to t_0} \frac{\mathbf{v}(t) - \mathbf{v}(t_0)}{t - t_0} = \dot{\mathbf{v}}(t_0),$$

exists and is finite, thus

$$\lim_{t\to t_0} (\mathbf{v}(t) - \mathbf{v}(t_0)) = \lim_{t\to t_0} \frac{\mathbf{v}(t) - \mathbf{v}(t_0)}{t - t_0}(t - t_0) = \dot{\mathbf{v}}(t_0) \lim_{t\to t_0} (t - t_0) = 0,$$

which, indeed, corresponds to the continuity of $\mathbf{v}(t)$ in t_0.

Note 1
While all differentiable functions are continuous, the converse is not true, i.e. there are functions which are continuous but not differentiable. An example is $f(x) = |x|$, which is continuous everywhere but not differentiable at $x = 0$.

Fig. 1.2 Geometric view of vector function derivative

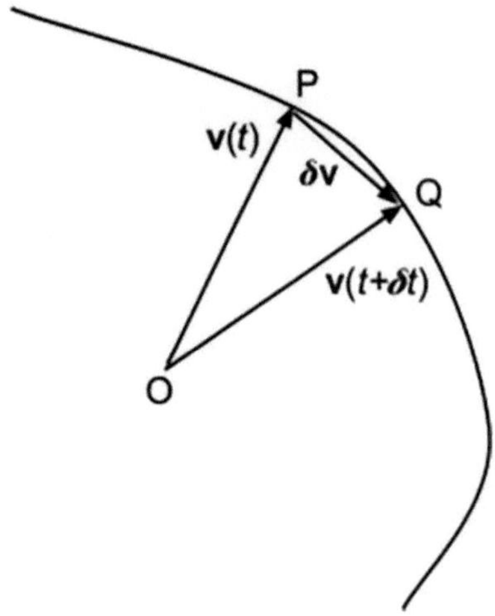

1.3.1 Geometric Interpretation

Let $\mathbf{v}(t)$ a continuous function and define

$$\delta\mathbf{v} = \mathbf{v}(t + \delta t) - \mathbf{v}(t), \tag{1.14}$$

which is a vector in the direction pointing from the point P to Q (Fig. 1.2).

If the limit

$$\lim_{\delta t \to 0} \frac{\delta\mathbf{v}}{\delta t} \equiv \frac{\mathrm{d}\mathbf{v}}{\mathrm{d}t} \tag{1.15}$$

is finite, we easily understand (see Fig. 1.2) that the vector derivative $\dfrac{\mathrm{d}v}{\mathrm{d}t}$ has the direction of the straight line tangent to the curve in P.

Exercise 1.1 Assuming $\mathbf{v}(t)$ to be a vector function of constant norm ($\|\mathbf{v}\| \equiv v =$ const.), show that it results $\mathbf{v} \cdot \dot{\mathbf{v}} = 0$ (i.e. $\dot{\mathbf{v}} \perp \mathbf{v}$).

Solution Actually, being $\|\mathbf{v}\|$ constant, we have

$$\frac{\mathrm{d}}{\mathrm{d}t}(\mathbf{v} \cdot \mathbf{v}) = 2\mathbf{v} \cdot \dot{\mathbf{v}} = \frac{\mathrm{d}}{\mathrm{d}t}\|\mathbf{v}\|^2 = 0.$$

Exercise 1.2 Given the vector functions $\mathbf{a}(t)$, $\mathbf{b}(t)$ and $\mathbf{c}(t)$, calculate the derivative $\dfrac{\mathrm{d}}{\mathrm{d}t}(\mathbf{a} \wedge \mathbf{b} \cdot \mathbf{c})$.

Solution

$$\frac{\mathrm{d}}{\mathrm{d}t}(\mathbf{a}\wedge\mathbf{b}\cdot\mathbf{c}) = \mathbf{c}\cdot\frac{\mathrm{d}}{\mathrm{d}t}(\mathbf{a}\wedge\mathbf{b}) + (\mathbf{a}\wedge\mathbf{b})\cdot\frac{\mathrm{d}\mathbf{c}}{\mathrm{d}t} =$$
$$= \mathbf{c}\cdot\left(\frac{\mathrm{d}\mathbf{a}}{\mathrm{d}t}\wedge\mathbf{b} + \mathbf{a}\wedge\frac{\mathrm{d}\mathbf{b}}{\mathrm{d}t}\right) + (\mathbf{a}\wedge\mathbf{b})\cdot\frac{\mathrm{d}\mathbf{c}}{\mathrm{d}t}.$$

Exercise 1.3 Calculate the derivative $\dfrac{\mathrm{d}}{\mathrm{d}t}(\mathbf{a}\wedge\mathbf{b}\wedge\mathbf{c})$.

Solution Left to the reader.

Exercise 1.4 Show that

$$\mathbf{e}_r \wedge \mathrm{d}\mathbf{e}_r = \frac{\mathbf{r}\wedge\mathrm{d}\mathbf{r}}{r^2},$$

where $\mathbf{e}_r \equiv \dfrac{\mathbf{r}}{r}$ is the unit radial vector.

Solution

$$\mathrm{d}\mathbf{e}_r = \mathrm{d}\left(\frac{\mathbf{r}}{r}\right) = \frac{1}{r}\,\mathrm{d}\mathbf{r} - \frac{\mathbf{r}}{r^2}\,\mathrm{d}r,$$

hence we have

$$\mathbf{e}_r \wedge \mathrm{d}\mathbf{e}_r = \mathbf{e}_r \wedge \left(\frac{1}{r}\,\mathrm{d}\mathbf{r} - \frac{\mathbf{r}}{r^2}\,\mathrm{d}r\right) = \frac{\mathbf{r}\wedge\mathrm{d}\mathbf{r}}{r^2}.$$

Moreover, because $\mathbf{e}_r$ is orthogonal to $\dot{\mathbf{e}}_r$ (i.e. $\widehat{\mathbf{e}_r\dot{\mathbf{e}}_r} = \frac{\pi}{2}$), it results

$$\|\mathbf{e}_r \wedge \dot{\mathbf{e}}_r\| = \|\dot{\mathbf{e}}_r\|.$$

Exercise 1.5 Show that $\mathbf{v}\wedge\dot{\mathbf{v}} = 0$ if and only if $\mathbf{v}$ has invariable direction ($\mathbf{e}_v = const.$).

Solution Geometrically, the solution is intuitive because the fixed direction of $\mathbf{v}$ (although v may vary) implies a variation $\delta\mathbf{v} \propto \mathbf{v}$, i.e. $\delta\mathbf{v} \parallel \mathbf{v}$ and so $\mathbf{v}\wedge\dot{\mathbf{v}} = 0$; vice versa, if $\mathbf{v}\wedge\dot{\mathbf{v}} = 0$ then $\delta\mathbf{v} \parallel \mathbf{v}$.

In a more rigorous way:

let $\mathbf{e}_v = \mathbf{v}/v$ the unit vector in the (fixed) direction of $\mathbf{v}$, then we can write $\mathbf{v} = v\,\mathbf{e}_v$, hence

$$\dot{\mathbf{v}} = \dot{v}\mathbf{e}_v + v\dot{\mathbf{e}}_v,$$

so that

$$\mathbf{v} \wedge \dot{\mathbf{v}} = v\mathbf{e}_v \wedge (\dot{v}\mathbf{e}_v + v\dot{\mathbf{e}}_v) = v^2\mathbf{e}_v \wedge \dot{\mathbf{e}}_v,$$

which implies

$$\mathbf{v} \wedge \dot{\mathbf{v}} = 0 \Longleftrightarrow \dot{\mathbf{e}_v} = 0 \Longleftrightarrow \mathbf{e}_v = const., \quad \text{or } \mathbf{e}_v \parallel \dot{\mathbf{v}}.$$

Exercise 1.6 Calculate the following

(*a*) $\dfrac{\mathrm{d}}{\mathrm{d}t}(\mathbf{r} \cdot \dot{\mathbf{r}})$,

(*b*) $\dfrac{\mathrm{d}}{\mathrm{d}t}(\mathbf{r} \wedge \dot{\mathbf{r}})$.

Solution

(*a*) $\dfrac{\mathrm{d}}{\mathrm{d}t}(\mathbf{r} \cdot \dot{\mathbf{r}}) = \dot{\mathbf{r}} \cdot \dot{\mathbf{r}} + \mathbf{r} \cdot \ddot{\mathbf{r}} = \|\dot{\mathbf{r}}\|^2 + \mathbf{r} \cdot \ddot{\mathbf{r}}$,

(*b*) $\dfrac{\mathrm{d}}{\mathrm{d}t}(\mathbf{r} \wedge \dot{\mathbf{r}}) = \dot{\mathbf{r}} \wedge \dot{\mathbf{r}} + \mathbf{r} \wedge \ddot{\mathbf{r}} = \mathbf{r} \wedge \ddot{\mathbf{r}}$.

1.4 Integration of Vector Functions

Definition 1.1 Let $\mathbf{f} = \mathbf{f}(t)$ a real vector function of the real variable $t \in [a, b]$. This function is said *integrable* over [*a*,*b*] if there exists a continuous function $\mathbf{F}(t)$ such that $\dfrac{\mathrm{d}\mathbf{F}}{\mathrm{d}t} = \mathbf{f}$, $\forall t \in [a, b]$. This formally translates into

$$\mathbf{F}(t) = \int_a^t \mathbf{f}(\tau)\,\mathrm{d}\tau, \quad t \in [a, b]. \tag{1.16}$$

Upon this definition, $\mathbf{F}(b) - \mathbf{F}(a) = [\mathbf{F}(x)]_a^b = \int_a^b \mathbf{f}(\tau)\,\mathrm{d}\tau$.[1] Of course, the above vector relation corresponds (in $\mathbb{R}^3$) to the following scalar equalities

$$\frac{\mathrm{d}F_x}{\mathrm{d}t} = f_x, \qquad \frac{\mathrm{d}F_y}{\mathrm{d}t} = f_y, \qquad \frac{\mathrm{d}F_z}{\mathrm{d}t} = f_z. \tag{1.17}$$

Exercise 1.7 Calculate the integral

$$\int \mathbf{r} \wedge \ddot{\mathbf{r}}\,\mathrm{d}t.$$

[1] As specified in Symbol section, $[F(x)]_a^b \equiv F(b) - F(a)$, definition used throughout this book.

Solution An integration by parts leads to

$$\int \mathbf{r}\wedge\ddot{\mathbf{r}}\,\mathrm{d}t = \int \mathbf{r}\wedge\frac{d\dot{\mathbf{r}}}{\mathrm{d}t}\,\mathrm{d}t = \int \mathbf{r}\wedge\mathrm{d}\dot{\mathbf{r}} = \mathbf{r}\wedge\dot{\mathbf{r}} - \int \dot{\mathbf{r}}\wedge\mathrm{d}\mathbf{r} = \mathbf{r}\wedge\dot{\mathbf{r}} + \mathbf{c},$$

given that $\dot{\mathbf{r}}\wedge\dot{\mathbf{r}} = 0$, and with $\mathbf{c}$ a constant vector. In the case of a particle subjected to a force (per unit mass) $\mathbf{F}$, we have $\mathbf{r}\wedge\ddot{\mathbf{r}} = \mathbf{r}\wedge\mathbf{F}$, which means that for a *central* force field ($\mathbf{r}\wedge\mathbf{F} = 0$) it results $\mathbf{r}\wedge\dot{\mathbf{r}} \equiv \mathbf{L} = \mathbf{c}'$, where $\mathbf{c}'$ is a constant vector which expresses the *integral* of angular momentum (per unit mass) $\mathbf{L}$.[2]

Exercise 1.8 Solve, with respect to $\mathbf{r}$, the equation

$$\mathbf{a}\wedge\ddot{\mathbf{r}} = \mathbf{b},$$

with $\mathbf{a}$ and $\mathbf{b}$ constant vectors.

Solution A double integration in time leads to

$$\mathbf{a}\wedge\dot{\mathbf{r}} = \mathbf{b}t + \mathbf{c}, \qquad \mathbf{a}\wedge\mathbf{r} = \frac{1}{2}\mathbf{b}t^2 + \mathbf{c}t + \mathbf{d},$$

the latter being an implicit expression for $\mathbf{r}$.

Exercise 1.9 Verify that $\mathbf{v}\cdot\dot{\mathbf{v}} = 0$ implies $\|\mathbf{v}\| \equiv v = \text{const.}$. Note that such result together with that of Ex. 1.1 constitutes a necessary and sufficient condition for $\mathbf{v}$ to have constant magnitude.

Solution Integrating both sides of the relation $\mathbf{v}\cdot\dot{\mathbf{v}} = 0$ gives

$$\int \mathbf{v}\cdot\dot{\mathbf{v}}\,\mathrm{d}t = \int \mathbf{v}\cdot\mathrm{d}\mathbf{v} = \frac{1}{2}\mathbf{v}\cdot\mathbf{v} = c,$$

where c is a constant, and so $\|\mathbf{v}\| = \sqrt{2c} = \text{const.}$

The equation above represents the *kinetic energy* theorem: when the acceleration $\mathbf{a} \equiv \dot{\mathbf{v}}$ is orthogonal to $\mathbf{v}$, then the unit mass particle kinetic energy, $T = \frac{1}{2}\mathbf{v}\cdot\mathbf{v}$, is constant. Note that the above relation is the inverse of that proved in Ex. 1.1, so that $\mathbf{v}\cdot\dot{\mathbf{v}} = 0$ and $\|\mathbf{v}\| = \text{const.}$ are equivalent.

1.5 The Formal Vector Operator ∇

The operator nabla, ∇, is defined as a (formal) differential vector

$$\nabla \equiv \mathbf{i}\frac{\partial}{\partial x} + \mathbf{j}\frac{\partial}{\partial y} + \mathbf{k}\frac{\partial}{\partial z}, \tag{1.18}$$

[2] An *integral* of motion is a function (either scalar or vector) which keeps constant along all possible solutions of the motion equations and which is independent of t.

in a 3D orthogonal Cartesian frame. As well known, the gradient of a scalar function $f(\mathbf{r})$ of class $C^1(\Omega)$, $\Omega \subseteq \mathbb{R}^n$, is a vector field whose components are the partial derivatives of the function. This formally translates into

$$\nabla f : \Omega \longrightarrow \mathbb{R}^n, \qquad \nabla f \equiv \text{grad} f = \mathbf{e}_i \frac{\partial f}{\partial x_i}, \tag{1.19}$$

where we adopted (as will be done in the following) the Einstein convention of implicit summation over a repeated index. By its definition, the vector operator ∇ is an application which maps a scalar field f of class $C^1(\Omega)$ to a continuous vector field ∇f.

Exercise 1.10 Show that $\nabla f(x, y, z)$ is a vector normal at the point (x, y, z) to the generic level surface of $f(x, y, z)$, $f(x, y, z) = c$, where c is an arbitrary real constant.

Solution As known, $\dfrac{\mathrm{d}\mathbf{r}}{\mathrm{d}s}$ is a vector tangent to a curve represented by $\mathbf{r} = \mathbf{r}(s)$, where s is a curvilinear abscissa. To solve the exercise it is sufficient to show that the orthogonality condition

$$\nabla f \cdot \frac{\mathrm{d}\mathbf{r}}{\mathrm{d}s} = 0$$

holds for an arbitrary curve belonging to the surface $f(x, y, z) = c$, passing through the point (x, y, z). This is easily seen because, indeed, on the level surface, where f is constant, it results

$$\nabla f \cdot \frac{\mathrm{d}\mathbf{r}}{\mathrm{d}s} = \frac{\partial f}{\partial x}\frac{\mathrm{d}x}{\mathrm{d}s} + \frac{\partial f}{\partial y}\frac{\mathrm{d}y}{\mathrm{d}s} + \frac{\partial f}{\partial z}\frac{\mathrm{d}z}{\mathrm{d}s} = \frac{\mathrm{d}f}{\mathrm{d}s} = 0.$$

1.5.1 Gradient and ∇ in Spherical Polar Coordinates

A point in $\mathbb{R}^3$ can be represented by 3 Cartesian coordinates, x, y, z, or other sets of 3 real numbers. One choice of coordinate system often advantageous to use is the system of spherical polar coordinates, (r, θ, φ) (see Fig. 1.3). In this frame, $r \geq 0$ and, for univocity, $0 \leq \theta < 2\pi, 0 \leq \varphi \leq \pi$.

The transformation from spherical polar to Cartesian coordinates is

$$\begin{cases} x = r \sin\varphi \cos\theta, \\ y = r \sin\varphi \sin\theta, \\ z = r \cos\varphi, \end{cases} \quad \text{with } r \geq 0,\ 0 \leq \theta < 2\pi,\ \text{and}\, 0 \leq \varphi \leq \pi, \tag{1.20}$$

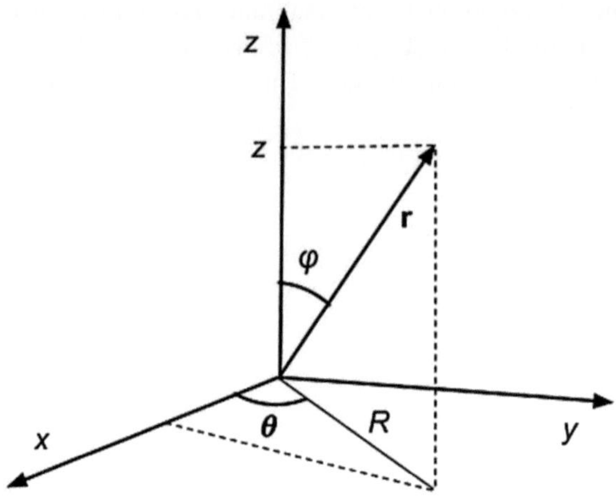

Fig. 1.3 Spherical polar (r, θ, φ) and cylindrical (R, θ, z) coordinates

which has the inverse transformation

$$\begin{cases} r = \sqrt{x^2 + y^2 + z^2}, \\ \theta = 2 \arctan \dfrac{y}{\sqrt{x^2 + y^2} + x}, \\ \varphi = \begin{cases} \arctan \dfrac{\sqrt{x^2 + y^2}}{z} & \text{for } z > 0, \\ \pi/2 & \text{for } z = 0, \\ \pi + \arctan \dfrac{\sqrt{x^2 + y^2}}{z} & \text{for } z < 0. \end{cases} \end{cases} \tag{1.21}$$

So, the spherical polar unit vectors are

$$\begin{aligned} \mathbf{e}_r &= \frac{\mathbf{r}}{r} = \sin\varphi \cos\theta\, \mathbf{i} + \sin\varphi \sin\theta\, \mathbf{j} + \cos\varphi\, \mathbf{k}, \\ \mathbf{e}_\theta &= \frac{\partial \mathbf{r}}{\partial \theta} \frac{1}{\dfrac{ds}{d\theta}} = -\sin\theta\, \mathbf{i} + \cos\theta\, \mathbf{j}, \\ \mathbf{e}_\varphi &= \frac{\partial \mathbf{r}}{\partial \varphi} \frac{1}{\dfrac{ds}{d\varphi}} = \cos\theta \cos\varphi\, \mathbf{i} + \cos\varphi \sin\theta\, \mathbf{j} - \sin\varphi\, \mathbf{k}, \end{aligned} \tag{1.22}$$

where we have made use of the relation $ds^2 = dx^2 + dy^2 + dz^2$, so that

$$\frac{\mathrm{d}s}{\mathrm{d}\theta} = \sqrt{\left(\frac{\mathrm{d}x}{\mathrm{d}\theta}\right)^2 + \left(\frac{\mathrm{d}y}{\mathrm{d}\theta}\right)^2 + \left(\frac{\mathrm{d}z}{\mathrm{d}\theta}\right)^2}, \tag{1.23}$$

and analogously for $\dfrac{\mathrm{d}s}{\mathrm{d}\varphi}$.

The expression of a generic vector $\mathbf{v}$ on the spherical polar basis is so $\mathbf{v} = (\mathbf{v} \cdot \mathbf{e}_r)\,\mathbf{e}_r + (\mathbf{v} \cdot \mathbf{e}_\theta)\,\mathbf{e}_\theta + (\mathbf{v} \cdot \mathbf{e}_\varphi)\,\mathbf{e}_\varphi \equiv v_r\,\mathbf{e}_r + v_\theta\,\mathbf{e}_\theta + v_\varphi\,\mathbf{e}_\varphi$.

The gradient operator of a function can be formally written as

$$\nabla f = (\nabla f \cdot \mathbf{e}_r)\,\mathbf{e}_r + (\nabla f \cdot \mathbf{e}_\theta)\,\mathbf{e}_\theta + (\nabla f \cdot \mathbf{e}_\varphi)\,\mathbf{e}_\varphi, \tag{1.24}$$

so as its explicit expression requires writing the polar components $\nabla f \cdot \mathbf{e}_r$, $\nabla f \cdot \mathbf{e}_\theta$ and $\nabla f \cdot \mathbf{e}_\varphi$ in terms of spherical polar coordinates. This is done taking into account that

$$\begin{cases} \nabla f \cdot \mathbf{e}_r = \dfrac{\partial f}{\partial x} \sin\varphi \cos\theta + \dfrac{\partial f}{\partial y} \sin\varphi \sin\theta + \dfrac{\partial f}{\partial z} \cos\varphi, \\[2ex] \nabla f \cdot \mathbf{e}_\theta = -\dfrac{\partial f}{\partial x} \sin\theta + \dfrac{\partial f}{\partial y} \cos\theta, \\[2ex] \nabla f \cdot \mathbf{e}_\varphi = \dfrac{\partial f}{\partial x} \cos\theta \cos\varphi + \dfrac{\partial f}{\partial y} \cos\varphi \sin\theta - \dfrac{\partial f}{\partial z} \sin\varphi, \end{cases} \tag{1.25}$$

and that

$$\begin{cases} \dfrac{\partial f}{\partial r} = \dfrac{\partial f}{\partial x}\dfrac{\partial x}{\partial r} + \dfrac{\partial f}{\partial y}\dfrac{\partial y}{\partial r} + \dfrac{\partial f}{\partial z}\dfrac{\partial z}{\partial r} = \dfrac{\partial f}{\partial x} \sin\varphi \cos\theta + \dfrac{\partial f}{\partial y} \sin\varphi \sin\theta + \dfrac{\partial f}{\partial z} \cos\varphi, \\[2ex] \dfrac{\partial f}{\partial \theta} = \dfrac{\partial f}{\partial x}\dfrac{\partial x}{\partial \theta} + \dfrac{\partial f}{\partial y}\dfrac{\partial y}{\partial \theta} + \dfrac{\partial f}{\partial z}\dfrac{\partial z}{\partial \theta} = -\dfrac{\partial f}{\partial x} r \sin\varphi \sin\theta + \dfrac{\partial f}{\partial y} r \sin\varphi \cos\theta, \\[2ex] \dfrac{\partial f}{\partial \varphi} = \dfrac{\partial f}{\partial x}\dfrac{\partial x}{\partial \varphi} + \dfrac{\partial f}{\partial y}\dfrac{\partial y}{\partial \varphi} + \dfrac{\partial f}{\partial z}\dfrac{\partial z}{\partial \varphi} = \dfrac{\partial f}{\partial x} r \cos\varphi \cos\theta + \dfrac{\partial f}{\partial y} r \cos\varphi \sin\theta - \dfrac{\partial f}{\partial z} r \sin\varphi. \end{cases} \tag{1.26}$$

To obtain the complete expression of the spherical polar components of the gradient of the function f, it is necessary, first, to solve for $\dfrac{\partial f}{\partial x}, \dfrac{\partial f}{\partial y}, \dfrac{\partial f}{\partial z}$ the system of equations 1.26. The system 1.26 is of the type

$$\begin{pmatrix} \dfrac{\partial f}{\partial r} \\[2ex] \dfrac{\partial f}{\partial \theta} \\[2ex] \dfrac{\partial f}{\partial \varphi} \end{pmatrix} = \underline{\mathbf{J}}^T \begin{pmatrix} \dfrac{\partial f}{\partial x} \\[2ex] \dfrac{\partial f}{\partial y} \\[2ex] \dfrac{\partial f}{\partial z} \end{pmatrix}, \tag{1.27}$$

where $\underline{\mathbf{J}}^T$ is the transposed of the Jacobian matrix of the Cartesian–spherical polar coordinate transformation

$$\underline{\mathbf{J}} = \frac{\partial(x, y, z)}{\partial(r, \theta, \varphi)} = \begin{pmatrix} \frac{\partial x}{\partial r} & \frac{\partial x}{\partial \theta} & \frac{\partial x}{\partial \varphi} \\ \frac{\partial y}{\partial r} & \frac{\partial y}{\partial \theta} & \frac{\partial y}{\partial \varphi} \\ \frac{\partial z}{\partial r} & \frac{\partial z}{\partial \theta} & \frac{\partial z}{\partial \varphi} \end{pmatrix}. \tag{1.28}$$

The formal solution of the problem is thus given by expressions 1.25, where the partial derivatives $\frac{\partial f}{\partial x}, \frac{\partial f}{\partial y}, \frac{\partial f}{\partial z}$ are functions of r, θ, φ through the solution of the system 1.26, which, formally, is

$$\begin{pmatrix} \frac{\partial f}{\partial x} \\ \frac{\partial f}{\partial y} \\ \frac{\partial f}{\partial z} \end{pmatrix} = \left(\underline{\mathbf{J}}^T\right)^{-1} \begin{pmatrix} \frac{\partial f}{\partial r} \\ \frac{\partial f}{\partial \theta} \\ \frac{\partial f}{\partial \varphi} \end{pmatrix}. \tag{1.29}$$

The general relation $\left(\underline{\mathbf{J}}^T\right)^{-1} = \left(\underline{\mathbf{J}}^{-1}\right)^T$ allows to obtain the solution of the above system by the knowledge of $\underline{\mathbf{J}}^{-1}$. In this case, $\underline{\mathbf{J}}^T$ is orthogonal so that $\left(\underline{\mathbf{J}}^T\right)^{-1} = \underline{\mathbf{J}}$, to insert in Eq.1.29 to obtain $\frac{\partial f}{\partial x}, \frac{\partial f}{\partial y}, \frac{\partial f}{\partial z}$ that, once inserted in Eq. 1.25, lead finally to

$$\begin{aligned} \nabla f \cdot \mathbf{e}_r &= \frac{\partial f}{\partial r}, \\ \nabla f \cdot \mathbf{e}_\theta &= \frac{1}{r \sin\varphi} \frac{\partial f}{\partial \theta}, \\ \nabla f \cdot \mathbf{e}_\varphi &= \frac{1}{r} \frac{\partial f}{\partial \varphi}. \end{aligned} \tag{1.30}$$

Note that the result above could have been more easily obtained by a straight comparison of Eqs. 1.25 and 1.26.

Finally, the gradient of the function f in spherical polars is thus

$$\nabla f = \frac{\partial f}{\partial r} \mathbf{e}_r + \frac{1}{r \sin\varphi} \frac{\partial f}{\partial \theta} \mathbf{e}_\theta + \frac{1}{r} \frac{\partial f}{\partial \varphi} \mathbf{e}_\varphi. \tag{1.31}$$

That means that in spherical polar coordinates the formal vector ∇ is written as

$$\nabla = \mathbf{e}_r \frac{\partial}{\partial r} + \mathbf{e}_\theta \frac{1}{r \sin\varphi} \frac{\partial}{\partial \theta} + \mathbf{e}_\varphi \frac{1}{r} \frac{\partial}{\partial \varphi}. \tag{1.32}$$

Note 2
In some textbooks the variables θ and φ are swapped with respect to our definition, so that

$$\nabla f = \frac{\partial f}{\partial r} \mathbf{e}_r + \frac{1}{r} \frac{\partial f}{\partial \theta} \mathbf{e}_\theta + \frac{1}{r \sin\theta} \frac{\partial f}{\partial \varphi} \mathbf{e}_\varphi. \tag{1.33}$$

1.5.2 Gradient and ∇ in Cylindrical Coordinates

Let us now express the gradient of a function in cylindrical coordinates.

Let (R, θ, z) be the cylindrical coordinates, graphically identified by Fig. 1.3, whose relation with Cartesian coordinates is

$$\begin{cases} x = R\cos\theta, \\ y = R\sin\theta, \qquad \text{with} \qquad 0 \le \theta < 2\pi \\ z = z. \end{cases} \tag{1.34}$$

The inverse (from cylindrical to Cartesian coordinates) transformation is simply $\theta = \text{atan2}(y, x)$, $R = \sqrt{x^2 + y^2}$, $z = z$, where $\text{atan2}(y, x)$ is the arctan function adapted to identify uniquely the θ angle between the x-axis and a ray from the origin to any (x, y) point in the Cartesian plane.

The system basis unit vectors are

$$\begin{cases} \mathbf{e}_R = \cos\theta\, \mathbf{i} + \sin\theta\, \mathbf{j}, \\ \mathbf{e}_\theta = \dfrac{\partial \mathbf{r}}{\partial \theta} \dfrac{1}{\frac{\mathrm{d}s}{\mathrm{d}\theta}} = -\sin\theta\, \mathbf{i} + \cos\theta\, \mathbf{j}, \\ \mathbf{e}_z = \mathbf{k}. \end{cases} \tag{1.35}$$

In such a basis, the generic vector $\mathbf{v}$ reads as $\mathbf{v} = (\mathbf{v} \cdot \mathbf{e}_R)\, \mathbf{e}_R + (\mathbf{v} \cdot \mathbf{e}_\theta)\, \mathbf{e}_\theta + (\mathbf{v} \cdot \mathbf{e}_z)\mathbf{e}_z \equiv v_R\, \mathbf{e}_R + v_\theta\, \mathbf{e}_\theta + v_z\, \mathbf{e}_z$. In such a basis, the gradient of a function $f(\mathbf{r})$ is written as

$$\nabla f = (\nabla f \cdot \mathbf{e}_R)\, \mathbf{e}_R + (\nabla f \cdot \mathbf{e}_\theta)\, \mathbf{e}_\theta + (\nabla f \cdot \mathbf{e}_z)\, \mathbf{e}_z, \tag{1.36}$$

where

$$\begin{cases} \nabla f \cdot \mathbf{e}_R = \dfrac{\partial f}{\partial x}\cos\theta + \dfrac{\partial f}{\partial y}\sin\theta, \\[2ex] \nabla f \cdot \mathbf{e}_\theta = -\dfrac{\partial f}{\partial x}\sin\theta + \dfrac{\partial f}{\partial y}\cos\theta, \\[2ex] \nabla f \cdot \mathbf{e}_z = \dfrac{\partial f}{\partial z}. \end{cases} \tag{1.37}$$

The partial derivatives of f with respect to R, θ, z are written as

$$\begin{cases} \dfrac{\partial f}{\partial R} = \dfrac{\partial f}{\partial x}\dfrac{\partial x}{\partial R} + \dfrac{\partial f}{\partial y}\dfrac{\partial y}{\partial R} + \dfrac{\partial f}{\partial z}\dfrac{\partial z}{\partial R} = \dfrac{\partial f}{\partial x}\cos\theta + \dfrac{\partial f}{\partial y}\sin\theta, \\[2ex] \dfrac{\partial f}{\partial \theta} = \dfrac{\partial f}{\partial x}\dfrac{\partial x}{\partial \theta} + \dfrac{\partial f}{\partial y}\dfrac{\partial y}{\partial \theta} + \dfrac{\partial f}{\partial z}\dfrac{\partial z}{\partial \theta} = -\dfrac{\partial f}{\partial x}\sin\theta + \dfrac{\partial f}{\partial y}\cos\theta, \\[2ex] \dfrac{\partial f}{\partial z} = \dfrac{\partial f}{\partial x}\dfrac{\partial x}{\partial z} + \dfrac{\partial f}{\partial y}\dfrac{\partial y}{\partial z} + \dfrac{\partial f}{\partial z}\dfrac{\partial z}{\partial z} = \dfrac{\partial f}{\partial z}, \end{cases} \tag{1.38}$$

and the general procedure to obtain the R, θ, z components of the gradient vector is the same as in the previous subsection. Also in this case, the comparison of (1.37) and (1.38) allows a straightforward determination of the cylindrical components of ∇f, giving

$$\nabla f = \frac{\partial f}{\partial R}\mathbf{e}_R + \frac{1}{R}\frac{\partial f}{\partial \theta}\mathbf{e}_\theta + \frac{\partial f}{\partial z}\mathbf{e}_z. \tag{1.39}$$

Finally, in cylindrical coordinates the formal vector ∇ is written as

$$\nabla = \mathbf{e}_R\,\frac{\partial}{\partial R} + \mathbf{e}_\theta\,\frac{1}{R}\frac{\partial}{\partial \theta} + \mathbf{e}_z\,\frac{\partial}{\partial z}. \tag{1.40}$$

1.6 The Divergence Operator

Given a differentiable vector field $\mathbf{v} = \mathbf{v}(\mathbf{r})$ with both $\mathbf{r}$ and $\mathbf{v}$ in $\mathbb{R}^n$, we call *divergence* of the vector $\mathbf{v}$ the sum of the partial derivatives of the vector components respect to homonymous coordinates

$$\operatorname{div}\mathbf{v} : C \longrightarrow \mathbb{R} \qquad\qquad \operatorname{div}\mathbf{v} \equiv \nabla\cdot\mathbf{v} = \frac{\partial v_i}{\partial x_i}, \tag{1.41}$$

where, again, we adopted Einstein summation convention.

Clearly, the divergence operator transforms a vector field into a scalar one. Formally, div $\mathbf{v}$ is expressible as the scalar product of ∇ with $\mathbf{v}$, $\nabla \cdot \mathbf{v}$.

A vector field with null divergence, $\nabla \cdot \mathbf{v} = 0$, is said *solenoidal*. As we will see in the following, the gravitational field generated by a regular mass distribution is solenoidal in the (empty) space external to the mass distribution.

Exercise 1.11 Letting $\mathbf{r} \in \mathbb{R}^n$, show that

$$\nabla \cdot \mathbf{r} = n. \tag{1.42}$$

Solution We have

$$\nabla \cdot \mathbf{r} = \frac{\partial x_i}{\partial x_i} = \underbrace{1 + 1 + \cdots + 1}_{n\, terms} = n.$$

Exercise 1.12 For $\mathbf{r} \in \mathbb{R}^n$, show that

$$\nabla \cdot \mathbf{e}_r = \frac{n-1}{r}. \tag{1.43}$$

Solution Adopting Einstein summation convention

$$\nabla \cdot \mathbf{e}_r = \frac{\partial}{\partial x_i}\frac{x_i}{r} = \frac{nr - \sum\limits_{i=1}^{n}(x_i^2/r)}{r^2} = \frac{nr - r}{r^2} = \frac{n-1}{r}.$$

The above result implies that, in $\mathbb{R}^3$, $\nabla \cdot \mathbf{e}_r = 2/r$.

Exercise 1.13 Let $f : \mathbb{R}^n \longrightarrow \mathbb{R}$ represent a scalar field and $\mathbf{v} : \mathbb{R}^n \longrightarrow \mathbb{R}^n$ a vector field. Show that the following relation holds:

$$\nabla \cdot f\mathbf{v} = f\nabla \cdot \mathbf{v} + \mathbf{v} \cdot \nabla f. \tag{1.44}$$

Solution

$$\nabla \cdot f\mathbf{v} = \frac{\partial}{\partial x_i}(f\, v_i) = f\,\frac{\partial v_i}{\partial x_i} + v_i\frac{\partial f}{\partial x_i}.$$

Note that by means of (1.44), the result (1.43) in Ex. 1.12 is easily obtained, because it is

$$\nabla \cdot \mathbf{e}_r = \nabla \cdot \frac{\mathbf{r}}{r} = \frac{1}{r}\nabla \cdot \mathbf{r} + \mathbf{r} \cdot \nabla\frac{1}{r} = \frac{n}{r} - \mathbf{r} \cdot \frac{\mathbf{r}}{r^2} = \frac{n}{r} - \frac{1}{r} = \frac{n-1}{r}.$$

Exercise 1.14 Calculate in $\mathbb{R}^3$

$$\nabla \cdot (\mathbf{r} \wedge \mathbf{a}), \quad \text{with} \quad \mathbf{a} = \text{const.}.$$

Solution

$$\nabla \cdot (\mathbf{r} \wedge \mathbf{a}) = \frac{\partial}{\partial x}(ya_z - za_y) - \frac{\partial}{\partial y}(xa_z - za_x) + \frac{\partial}{\partial z}(xa_y - ya_x) = 0.$$

Such result is obtained also by applying the rules of the scalar triple product to the formal vector ∇, $\mathbf{r}$ and $\mathbf{a}$.[3]

Exercise 1.15 Show that in $\mathbb{R}^n$

$$\nabla \cdot \mathbf{v} = \nabla v_i \cdot \mathbf{e}_i .$$

Solution

$$\nabla v_i \cdot \mathbf{e}_i = \mathbf{e}_k \frac{\partial v_i}{\partial x_k} \cdot \mathbf{e}_i = \frac{\partial v_i}{\partial x_i} = \nabla \cdot \mathbf{v}.$$

Exercise 1.16 In $\mathbb{R}^n$, given the scalar function $\varphi(\mathbf{r})$, calculate in a direct way the divergence of the gradient of φ, i.e.

$$\nabla \cdot \nabla\varphi \equiv (\nabla \cdot \nabla)\varphi = \nabla^2\varphi, \tag{1.45}$$

where $\nabla^2 \equiv \sum_{i=1}^{n} \frac{\partial^2}{\partial x_i^2}$ is the *Laplacian* formal operator, often written also as Δ.

Solution

$$\nabla \cdot \nabla\varphi = \mathbf{e}_i \frac{\partial}{\partial x_i} \cdot \frac{\partial \varphi}{\partial x_i}\mathbf{e}_i = \sum_{i=1}^{n} \frac{\partial^2 \varphi}{\partial x_i^2} = \nabla^2\varphi \equiv \Delta\varphi.$$

Exercise 1.17 Compute the gradient of the divergence of $\mathbf{e}_r$, $\nabla\, \nabla \cdot \mathbf{e}_r$.

Solution

$$\nabla\, \nabla \cdot \mathbf{e}_r = \nabla \frac{n-1}{r} = (n-1)\nabla \frac{1}{r} = \frac{1-n}{r^2}\nabla r = \frac{1-n}{r^2}\,\mathbf{e}_r = (1-n)\frac{\mathbf{r}}{r^2}.$$

Exercise 1.18 Given $\mathbf{r} \in \mathbb{R}^3$ and a differentiable function $f(r)$, show that

$$\nabla \cdot f\mathbf{e}_r = \frac{1}{r^2}\frac{\mathrm{d}}{\mathrm{d}r}\left(r^2 f\right). \tag{1.46}$$

Solution In $\mathbb{R}^n$, by means of Eq. 1.44 it results

[3] The scalar triple product of 3 vectors satisfies various identities, such as $\mathbf{a} \cdot (\mathbf{b} \wedge \mathbf{c}) = \det(\mathbf{a}, \mathbf{b}, \mathbf{c})$.

$$\nabla \cdot f\mathbf{e}_r = f\nabla \cdot \mathbf{e}_r + \mathbf{e}_r \cdot \nabla f = f\frac{n-1}{r} + \mathbf{e}_r \cdot f'(r)\mathbf{e}_r = \frac{1}{r^2}\left[(n-1)fr + r^2 f'\right],$$

where the prime, $'$, indicates derivative with respect to r. For $n = 3$ the expression above reduces to

$$\nabla \cdot f\mathbf{e}_r = \frac{1}{r^2}\left[2fr + r^2 f'\right] = \frac{1}{r^2}\frac{\mathrm{d}}{\mathrm{d}r}\left(r^2 f\right).$$

Let us note that $f(r)\,\mathbf{e}_r$ could represent a central force field, *attractive* in the region where $f(r) < 0$ and *repulsive* where $f(r) > 0$.

Exercise 1.19 In $\mathbb{R}^n$, calculate explicitly

(*i*) $\nabla \cdot \nabla \frac{1}{r}$,

(*ii*) $\nabla \cdot \nabla \ln \frac{1}{r}$.

Solution With regard to (*i*)

$$\begin{aligned} \nabla \cdot \nabla \frac{1}{r} &= \nabla \cdot \left(-\frac{1}{r^2}\nabla r\right) = \nabla \cdot \left(-\frac{1}{r^2}\frac{\mathbf{r}}{r}\right) = -\frac{1}{r^3}\nabla \cdot \mathbf{r} - \mathbf{r} \cdot \nabla \frac{1}{r^3} = \\ &= -\frac{1}{r^3}n + \mathbf{r} \cdot \frac{3}{r^4}\nabla r = -\frac{n}{r^3} + \frac{3}{r^4}\mathbf{r} \cdot \frac{\mathbf{r}}{r} = \frac{3-n}{r^3}, \end{aligned} \tag{1.47}$$

so that

$$\nabla \cdot \nabla \frac{1}{r} = \nabla^2 \frac{1}{r} = \frac{3-n}{r^3} = 0 \quad \Longleftrightarrow \quad n = 3. \tag{1.48}$$

This means that in $\mathbb{R}^3$ the function $1/r$ is *harmonic*, because its Laplacian is zero.[4]

With regard to (*ii*)

$$\begin{aligned} \nabla \cdot \nabla \ln \frac{1}{r} &= \nabla \cdot \left(-\frac{1}{r}\nabla r\right) = -\frac{1}{r^2}\nabla \cdot \mathbf{r} - \mathbf{r} \cdot \nabla \frac{1}{r^2} = \\ &= -\frac{n}{r^2} + 2\frac{\mathbf{r}}{r^3} \cdot \nabla r = \frac{2-n}{r^2}, \end{aligned} \tag{1.49}$$

so that

$$\nabla \cdot \nabla \ln \frac{1}{r} = \frac{2-n}{r^2} = 0 \quad \Longleftrightarrow \quad n = 2. \tag{1.50}$$

[4] A function is called harmonic if its Laplacian is zero in every point of its domain.

As a consequence, the function $\ln(1/r)$ (and also $\ln r$) is harmonic in $\mathbb{R}^2$ (it is the so-called *logarithmic* potential).

Exercise 1.20 Verify that in $\mathbb{R}^n$

$$\nabla \cdot \nabla \frac{1}{r^{n-2}} = 0, \quad \forall \,\text{integer}\, n \geq 0. \tag{1.51}$$

Solution Actually, it results that

$$\begin{aligned}\nabla \cdot \nabla \frac{1}{r^{n-2}} &= \nabla \cdot (2-n)\, r^{1-n} \nabla r = (2-n) \frac{r^{1-n}}{r} \nabla \cdot \mathbf{r} + \mathbf{r} \cdot \nabla (2-n) r^{-n} = \\ &= (2-n) n r^{-n} + (2-n) \mathbf{r} \cdot (-n) r^{-n-1} \nabla r = \\ &= (2-n) n r^{-n} + (2-n)(-n) r^{-n-1} \frac{\mathbf{r} \cdot \mathbf{r}}{r} = \\ &= (2-n) n r^{-n} + (2-n)(-n) r^{-n} = 0.\end{aligned}$$

In particular, if $n = 2$ then $\dfrac{1}{r^{n-2}} = 1$, which is a constant and so all its derivatives are zero. If $n = 1$, then $\dfrac{1}{r^{n-2}} = r = |x|$ and, clearly, every 1st degree polynomial in the single variable x, $p_1(x) = ax + b$, is such to give

$$\nabla \cdot \nabla p_1(x) = 0.$$

1.7 The Curl

The curl of a vector field $\mathbf{v}$ of class $C^1(\Omega)$, where $\Omega \in \mathbb{R}^3$, is also a vector field, written as $\text{curl}\, \mathbf{v} : \Omega \longrightarrow \mathbb{R}^3$, defined (in Cartesian coordinates) as

$$\text{curl}\, \mathbf{v} = \left(\frac{\partial v_z}{\partial y} - \frac{\partial v_y}{\partial z} \right) \mathbf{i} - \left(\frac{\partial v_z}{\partial x} - \frac{\partial v_x}{\partial z} \right) \mathbf{j} + \left(\frac{\partial v_y}{\partial x} - \frac{\partial v_x}{\partial y} \right) \mathbf{k}. \tag{1.52}$$

Similarly to the divergence, which can be written as the formal scalar product of ∇ and $\mathbf{v}$, the curl of $\mathbf{v}$ can be formally written as the vector product between ∇ and $\mathbf{v}$

$$\text{curl}\, \mathbf{v} = \nabla \wedge \mathbf{v} = \begin{vmatrix} \mathbf{i} & \mathbf{j} & \mathbf{k} \\ \dfrac{\partial}{\partial x} & \dfrac{\partial}{\partial y} & \dfrac{\partial}{\partial z} \\ v_x & v_y & v_z \end{vmatrix}. \tag{1.53}$$

Contrary to the divergence, the curl is defined only in $\mathbb{R}^2$ and $\mathbb{R}^3$.

Definition 1.2 A vector field $\mathbf{v} \in \mathcal{C}^1(\Omega)$ is called *irrotational* if

$$\nabla \wedge \mathbf{v} = 0. \tag{1.54}$$

Definition 1.3 A vector field $\mathbf{v}$ is called *conservative* if it can be written as gradient of a scalar function, i.e. if there exists a scalar field $f(\mathbf{r})$ of class $\mathcal{C}^1$ (the *potential* of $\mathbf{v}$) such that $\mathbf{v} = \nabla f$.

Lemma 1.1 *Let* $\mathbf{v}$ *be a conservative vector field of class* $\mathcal{C}^1$, *whose potential is a scalar field* $f(\mathbf{r})$ *of class* $\mathcal{C}^2$.

It results

$$\operatorname{curl} \mathbf{v} = \nabla \wedge \nabla f = 0. \tag{1.55}$$

The easy demonstration is left to the reader.

Corollary 1.1 *A conservative vector field of class* $\mathcal{C}^1(\Omega)$ *is irrotational. The vice versa, in general, is not true, unless* Ω *is a simply connected domain.*

Note 3

Note that throughout this book we assume the convention for a conservative force field and its associated potential, usually indicated by $\mathbf{F}$ and U, that $\mathbf{F} = \nabla U$, instead of the often used convention of assuming a $-$ sign in front of the potential. In this book, U is the *potential* and $V = -U$ is the *potential energy*. They are both specific quantities, i.e. per unit mass.

Exercise 1.21 Show that

$$\nabla \wedge \mathbf{r} = 0. \tag{1.56}$$

Solution

$$\operatorname{curl} \mathbf{r} = \nabla \wedge \mathbf{r} = \begin{vmatrix} \mathbf{i} & \mathbf{j} & \mathbf{k} \\ \dfrac{\partial}{\partial x} & \dfrac{\partial}{\partial y} & \dfrac{\partial}{\partial z} \\ x & y & z \end{vmatrix} = 0,$$

because every partial derivative $\dfrac{\partial x_i}{\partial x_j}$ with $i \neq j$ is identically zero.

Exercise 1.22 Given the vector function $\mathbf{F} : \mathbb{R}^3 \longrightarrow \mathbb{R}^3$ of class $\mathcal{C}^1$, show that

$$\nabla \wedge \mathbf{F} = 0 \quad \Longleftrightarrow \quad \mathbf{F} = \nabla\varphi, \tag{1.57}$$

where $\varphi(\mathbf{r})$ is a scalar function of class $\mathcal{C}^2$.

Solution ($\Longleftarrow$) If $\mathbf{F} = \nabla\varphi$, then the curl of $\mathbf{F}$ is

$$\nabla \wedge \mathbf{F} = \begin{vmatrix} \mathbf{i} & \mathbf{j} & \boldsymbol{k} \\ \dfrac{\partial}{\partial x} & \dfrac{\partial}{\partial y} & \dfrac{\partial}{\partial z} \\ \dfrac{\partial\varphi}{\partial x} & \dfrac{\partial\varphi}{\partial y} & \dfrac{\partial\varphi}{\partial z} \end{vmatrix} = (\varphi_{zy} - \varphi_{yz})\,\mathbf{i} - (\varphi_{xz} - \varphi_{zx})\,\mathbf{j} + (\varphi_{yx} - \varphi_{xy})\,\mathbf{k},$$

where $\varphi_{xy} \equiv \frac{\partial}{\partial x}\frac{\partial\varphi}{\partial y}$ and similarly for the other derivatives. The above curl vanishes because the function φ is of class $\mathcal{C}^2$ and so, by Schwarz's theorem, the order of partial derivation can be exchanged, i.e. $\varphi_{xy} = \varphi_{yx}$, etc..

($\Longrightarrow$) On the other side, if $\nabla \wedge \mathbf{F} = 0$ then

$$\nabla \wedge \mathbf{F} = \left(\frac{\partial F_z}{\partial y} - \frac{\partial F_y}{\partial z}\right)\mathbf{i} - \left(\frac{\partial F_z}{\partial x} - \frac{\partial F_x}{\partial z}\right)\mathbf{j} + \left(\frac{\partial F_y}{\partial x} - \frac{\partial F_x}{\partial y}\right)\mathbf{k} = 0,$$

that implies

$$\frac{\partial F_y}{\partial z} = \frac{\partial F_z}{\partial y}, \qquad \frac{\partial F_z}{\partial x} = \frac{\partial F_x}{\partial z}, \qquad \frac{\partial F_y}{\partial x} = \frac{\partial F_x}{\partial y}.$$

The expressions above represent indeed the conditions for the differential form $\mathbf{F} \cdot \mathrm{d}\mathbf{r} = F_x\mathrm{d}x + F_y\mathrm{d}y + F_z\mathrm{d}z$ to be an *exact* differential. This means that it does exist a continuous and differentiable function $\varphi(\mathbf{r})$ whose differential reproduces the given form, $\mathrm{d}\varphi = \dfrac{\partial\varphi}{\partial x}\mathrm{d}x + \dfrac{\partial\varphi}{\partial y}\mathrm{d}y + \dfrac{\partial\varphi}{\partial z}\mathrm{d}z = F_x\mathrm{d}x + F_y\mathrm{d}y + F_z\mathrm{d}z$, which is equivalent to

$$\mathbf{F} = \nabla\varphi.$$

In conclusion, in a simply connected domain $D \subseteq \mathbb{R}^3$, a vector field is conservative if and only if it is irrotational.

Exercise 1.23 Show that if $\mathbf{F}$ admits a potential then it results

$$\nabla \cdot (\mathbf{F} \wedge \mathbf{r}) = 0.$$

Solution Since

$$\mathbf{F} \wedge \mathbf{r} = (F_y z - F_z y)\,\mathbf{i} - (F_x z - F_z x)\,\mathbf{j} + (F_x y - F_y x)\,\mathbf{k},$$

it is

$$\nabla \cdot (\mathbf{F} \wedge \mathbf{r}) = \frac{\partial}{\partial x}(F_y z - F_z y) - \frac{\partial}{\partial y}(F_x z - F_z x) + \frac{\partial}{\partial z}(F_x y - F_y x) =$$
$$= \frac{\partial F_y}{\partial x} z - \frac{\partial F_z}{\partial x} y - \frac{\partial F_x}{\partial y} z + \frac{\partial F_z}{\partial y} x + \frac{\partial F_x}{\partial z} y - \frac{\partial F_y}{\partial z} x =$$
$$= \mathbf{r} \cdot \nabla \wedge \mathbf{F}.$$

In the hypothesis of this exercise it is $\nabla \wedge \mathbf{F} = 0$, from which it follows

$$\nabla \cdot \mathbf{F} \wedge \mathbf{r} = 0.$$

Exercise 1.24 Given a scalar field f and a vector field $\mathbf{v}$, show that

$$\nabla \wedge f\mathbf{v} = f\nabla \wedge \mathbf{v} - \mathbf{v} \wedge \nabla f = f\nabla \wedge \mathbf{v} + \nabla f \wedge \mathbf{v}. \tag{1.58}$$

Solution The solution is left to the reader. Note that the expression 1.58 is similar to that of the scalar product of ∇ with $f\mathbf{v}$ (Eq. 1.44) but for the minus sign.

Exercise 1.25 Determine the real values of a, b and c for which the field

$$\mathbf{F} = (x + 2y + az)\,\mathbf{i} + (bx - 3y - z)\,\mathbf{j} + (4x + cy - 2z)\,\mathbf{k}$$

is conservative, and find the potential function.

Solution It is easily found that $a = 4,\ b = 2,\ c = -1$, and that the potential is

$$U(x, y, z) = \frac{x^2}{2} + 2x(y + 2z) - \frac{3}{2}y\left(y + \frac{2}{3}z\right) - z^2 + const..$$

Exercise 1.26 Show that

$$\nabla \wedge (\nabla \wedge \mathbf{v}) = \nabla(\nabla \cdot \mathbf{v}) - \nabla^2\mathbf{v},$$

where $\nabla^2\mathbf{v}$ is the vector whose components are $\nabla^2 v_x$, $\nabla^2 v_y$, $\nabla^2 v_z$.

Solution By a straight formal application of the vector triple product rule[5] where we assume $\mathbf{a} = \nabla$, $\mathbf{b} = \nabla$, $\mathbf{c} = \mathbf{v}$, it results

$$\nabla \wedge (\nabla \wedge \mathbf{v}) = \nabla(\nabla \cdot \mathbf{v}) - (\nabla \cdot \nabla)\mathbf{v} = \nabla(\nabla \cdot \mathbf{v}) - \nabla^2\mathbf{v}.$$

---

[5] The vector triple product rule says that $\mathbf{a} \wedge (\mathbf{b} \wedge \mathbf{c}) = \mathbf{b}(\mathbf{a} \cdot \mathbf{c}) - \mathbf{c}(\mathbf{a} \cdot \mathbf{b})$.

A longer way to the same result is via a thorough development of the calculations

$$\nabla \wedge (\nabla \wedge \mathbf{v}) = -\left(\frac{\partial^2}{\partial x^2} + \frac{\partial^2}{\partial y^2} + \frac{\partial^2}{\partial z^2}\right)(v_x\mathbf{i} + v_y\mathbf{j} + v_z\mathbf{k})$$
$$+ \frac{\partial}{\partial x}\left(\frac{\partial v_x}{\partial x} + \frac{\partial v_y}{\partial y} + \frac{\partial v_z}{\partial z}\right)\mathbf{i} + +\frac{\partial}{\partial y}\left(\frac{\partial v_x}{\partial x} + \frac{\partial v_y}{\partial y} + \frac{\partial v_z}{\partial z}\right)\mathbf{j}$$
$$+ \frac{\partial}{\partial z}\left(\frac{\partial v_x}{\partial x} + \frac{\partial v_y}{\partial y} + \frac{\partial v_z}{\partial z}\right)\mathbf{k} = -\nabla^2\mathbf{v} + \nabla\nabla \cdot \mathbf{v}.$$

Exercise 1.27 Show that a central force field $\mathbf{F}(\mathbf{r}) = f(\mathbf{r})\,\mathbf{e}_r$ is conservative if, and only if, it is spherically symmetric, i.e. $f(\mathbf{r}) = f(r)$.

Solution Given a central force field $\mathbf{F}(\mathbf{r}) = f(\mathbf{r})\,\mathbf{e}_r$, it is

$$\nabla \wedge f\mathbf{e}_r = f\nabla \wedge \mathbf{e}_r - \mathbf{e}_r \wedge \nabla f,$$

and, being $\nabla \wedge \mathbf{e}_r = 0$, it follows

$$\nabla \wedge \mathbf{F} = \nabla \wedge f\mathbf{e}_r = -\mathbf{e}_r \wedge \nabla f.$$

Now

(*i*) If the field is conservative, i.e. $\nabla \wedge f\mathbf{e}_r = 0$, the above equation gives ∇f parallel to $\mathbf{e}_r$ i.e. f is a function of r;
(*ii*) if the field is spherically symmetric ($f(\mathbf{r}) = f(r)$) ∇f is (see Ex. 1.10) orthogonal to any spherical surface centred at the origin, which is its level surface, and thus ∇f is radially oriented and $\nabla \wedge f\mathbf{e}_r = 0$. This result is contained in the formulas

$$\nabla f(r) = f'(r)\,\mathbf{e}_r \quad \Longrightarrow \quad \mathbf{e}_r \wedge \nabla f = 0.$$

1.8 Formal Calculation of Divergence and Laplacian by the Operator ∇

Using the properties satisfied by the formal vector ∇, we reobtain here the expressions of the divergence and Laplacian of vector and scalar fields.

1.8.1 Spherical Polar Coordinates

Given the formal vector ∇ written as (Eq. 1.32)

$$\nabla = \mathbf{e}_r \frac{\partial}{\partial r} + \mathbf{e}_\theta \frac{1}{r \sin\varphi} \frac{\partial}{\partial \theta} + \mathbf{e}_\varphi \frac{1}{r} \frac{\partial}{\partial \varphi}, \tag{1.59}$$

and given a generic vector field $\mathbf{F}$ in spherical polar coordinates

$$\mathbf{F} = F_r \mathbf{e}_r + F_\theta \mathbf{e}_\theta + F_\varphi \mathbf{e}_\varphi, \tag{1.60}$$

its divergence is formally expressed as

$$\begin{aligned} \nabla \cdot \mathbf{F} &= \left(\mathbf{e}_r \frac{\partial}{\partial r} + \frac{1}{r \sin\varphi} \mathbf{e}_\theta \frac{\partial}{\partial \theta} + \frac{1}{r} \mathbf{e}_\varphi \frac{\partial}{\partial \varphi} \right) \cdot \mathbf{F} = \\ &= \mathbf{e}_r \cdot \frac{\partial \mathbf{F}}{\partial r} + \frac{1}{r \sin\varphi} \mathbf{e}_\theta \cdot \frac{\partial \mathbf{F}}{\partial \theta} + \frac{1}{r} \mathbf{e}_\varphi \cdot \frac{\partial \mathbf{F}}{\partial \varphi}. \end{aligned} \tag{1.61}$$

Given the expressions 1.22, it results that

$$\frac{\partial \mathbf{e}_r}{\partial r} = \frac{\partial \mathbf{e}_\theta}{\partial r} = \frac{\partial \mathbf{e}_\theta}{\partial \varphi} = \frac{\partial \mathbf{e}_\varphi}{\partial r} = 0. \tag{1.62}$$

Moreover

$$\begin{aligned} \frac{\partial \mathbf{e}_r}{\partial \theta} &= -\sin\varphi \sin\theta \, \mathbf{i} + \sin\varphi \cos\theta \, \mathbf{j} = \sin\varphi \, \mathbf{e}_\theta, \\ \frac{\partial \mathbf{e}_r}{\partial \varphi} &= \cos\varphi \cos\theta \, \mathbf{i} + \cos\varphi \sin\theta \, \mathbf{j} - \sin\varphi \boldsymbol{k} = \mathbf{e}_\varphi, \\ \frac{\partial \mathbf{e}_\theta}{\partial \theta} &= -\cos\theta \, \mathbf{i} - \sin\theta \, \boldsymbol{j}, \\ \frac{\partial \mathbf{e}_\varphi}{\partial \theta} &= -\sin\theta \cos\varphi \, \mathbf{i} + \cos\varphi \cos\theta \, \mathbf{j} = \cos\varphi \, \mathbf{e}_\theta, \\ \frac{\partial \mathbf{e}_\varphi}{\partial \varphi} &= -\cos\theta \sin\varphi \, \mathbf{i} - \sin\varphi \sin\theta \, \mathbf{j} - \cos\varphi \, \mathbf{k} = -\mathbf{e}_r, \end{aligned} \tag{1.63}$$

which imply for the scalar products in Eq. 1.61 the expression

$$\begin{aligned} \nabla \cdot \mathbf{F} &= \frac{\partial F_r}{\partial r} + \frac{1}{r \sin\varphi} \left(F_r \sin\varphi + \frac{\partial F_\theta}{\partial \theta} + \cos\varphi \, F_\varphi \right) + \frac{1}{r} \left(F_r + \frac{\partial F_\varphi}{\partial \varphi} \right) = \\ &= \frac{\partial F_r}{\partial r} + \frac{2}{r} F_r + \frac{1}{r \sin\varphi} \frac{\partial F_\theta}{\partial \theta} + \frac{1}{r \sin\varphi} \left(\cos\varphi \, F_\varphi + \sin\varphi \frac{\partial F_\varphi}{\partial \varphi} \right) = \\ &= \frac{1}{r^2} \frac{\partial}{\partial r} \left(r^2 F_r \right) + \frac{1}{r \sin\varphi} \frac{\partial F_\theta}{\partial \theta} + \frac{1}{r \sin\varphi} \frac{\partial}{\partial \varphi} \left(\sin\varphi \, F_\varphi \right). \end{aligned} \tag{1.64}$$

Finally, the Laplacian of a scalar function $U(\mathbf{r})$ in spherical polars is obtained simply by letting $\mathbf{F} = \nabla U$ in the above expression, to obtain

$$\begin{aligned}\nabla^2 U \equiv \nabla\cdot\nabla U &= \frac{1}{r^2}\frac{\partial}{\partial r}\left(r^2\frac{\partial U}{\partial r}\right)+\frac{1}{r^2\sin^2\varphi}\frac{\partial^2 U}{\partial\theta^2}+\frac{1}{r\sin\varphi}\frac{\partial}{\partial\varphi}\left(\frac{1}{r}\sin\varphi\frac{\partial U}{\partial\varphi}\right)= \\ &= \frac{\partial^2 U}{\partial r^2}+\frac{2}{r}\frac{\partial U}{\partial r}+\frac{1}{r^2\sin^2\varphi}\frac{\partial^2 U}{\partial\theta^2}+\frac{1}{r^2\sin\varphi}\frac{\partial}{\partial\varphi}\left(\sin\varphi\frac{\partial U}{\partial\varphi}\right).\end{aligned} \tag{1.65}$$

1.8.2 Cylindrical Coordinates

In such a coordinate system the formal vector ∇ is written as (Eq. 1.40)

$$\nabla = \mathbf{e}_R\frac{\partial}{\partial R}+\mathbf{e}_\theta\frac{1}{R}\frac{\partial}{\partial\theta}+\mathbf{e}_z\frac{\partial}{\partial z}, \tag{1.66}$$

and the vector $\mathbf{F}$ on the cylindrical basis is

$$\mathbf{F} = F_R\,\mathbf{e}_R + F_\theta\,\mathbf{e}_\theta + F_z\,\mathbf{e}_z. \tag{1.67}$$

Consequently,

$$\nabla\cdot\mathbf{F} = \left(\mathbf{e}_R\frac{\partial}{\partial R}+\mathbf{e}_\theta\frac{1}{R}\frac{\partial}{\partial\theta}+\mathbf{e}_z\frac{\partial}{\partial z}\right)\cdot\mathbf{F}. \tag{1.68}$$

Now, given the orthonormality of the cylindrical vector basis, and given the expressions 1.35, we have

$$\frac{\partial\mathbf{e}_R}{\partial R}=\frac{\partial\mathbf{e}_R}{\partial z}=\frac{\partial\mathbf{e}_\theta}{\partial R}=\frac{\partial\mathbf{e}_\theta}{\partial z}=\frac{\partial\mathbf{e}_z}{\partial R}=\frac{\partial\mathbf{e}_z}{\partial\theta}=\frac{\partial\mathbf{e}_z}{\partial z}=0, \tag{1.69}$$

and

$$\begin{aligned}\frac{\partial\mathbf{e}_R}{\partial\theta} &= -\sin\theta\,\mathbf{i}+\cos\theta\,\mathbf{j}=\mathbf{e}_\theta,\\ \frac{\partial\mathbf{e}_\theta}{\partial\theta} &= -\cos\theta\,\mathbf{i}-\sin\theta\,\mathbf{j}=-\mathbf{e}_R,\end{aligned} \tag{1.70}$$

so that Eq. 1.68 becomes

$$\nabla\cdot\mathbf{F} = \frac{1}{R}\frac{\partial}{\partial R}(RF_R)+\frac{1}{R}\frac{\partial F_\theta}{\partial\theta}+\frac{\partial F_z}{\partial z}, \tag{1.71}$$

and, so, the Laplacian of a scalar function $U(\mathbf{r})$ in cylindrical coordinates is

$$\nabla^2 U \equiv \nabla \cdot \nabla U = \frac{1}{R}\frac{\partial}{\partial R}\left(R\frac{\partial U}{\partial R}\right) + \frac{1}{R^2}\frac{\partial^2 U}{\partial \theta^2} + \frac{\partial^2 U}{\partial z^2}. \tag{1.72}$$

1.9 Vector Fields

In $\mathbb{R}^n$, a generic vector field is a function of n real variables $(x_1, x_2, ..., x_n)$ which can be written on a basis $\{\mathbf{e}_i\}$ as $\mathbf{F}(\mathbf{r}) = F_i \mathbf{e}_i$.

For $n = 3$, it is

$$\mathbf{F}(x, y, z) = F_x\,\mathbf{i} + F_y\,\mathbf{j} + F_z\,\mathbf{k}. \tag{1.73}$$

A typical example is that of the *force* field generated by a charged particle (Coulombian "charges"). In this case the force is radially directed towards the particle (source of the field) if the test particle in the generic position $\mathbf{r}$ has a charge of opposite sign, or in the outward side for same sign. The case of a single particle which exerts a force is a particular case of *central* force field. This central force field has, obviously, a spherically symmetric *intensity*, i.e. $\mathbf{F} = f(r)\,\mathbf{e}_r$.

Of course, the sign of the function $f(r)$ discriminates between the *attractive* and *repulsive* case. The scaling of $f(r)$ with r, in the gravitational and Coulombian cases, is

$$\begin{cases} f(r) = -\dfrac{a^2}{r^2} < 0, & \text{attractive force,} \\ f(r) = \dfrac{a^2}{r^2} > 0, & \text{repulsive force,} \end{cases} \tag{1.74}$$

where the constant a^2 accounts for the coupling physical constant. In the Newtonian case, it is $a^2 = Gm$, where G is the universal gravitational constant and m the mass situated at the origin and generating the field. The gravitational force is always attractive.

1.9.1 Lines of Force of a Vector Field

Definition 1.4 The *lines of force* (or, more generally, *field lines*) of a vector field $\mathbf{F}$ are curves in $\mathbb{R}^n$ whose tangent at every point is in the direction of the field. As we will see, an information on the *intensity* of the field is possible for solenoidal fields, and it is given by the density of the lines of force (see Sect. 1.14).

Mathematically, the lines of force are represented by the curves $\mathbf{r}(s)$ (s is the curvilinear abscissa along the curve) such that

$$\frac{d\mathbf{r}}{ds} = c\mathbf{F}(\mathbf{r}), \tag{1.75}$$

where, for every arbitrary value of the non-zero constant c, $\dfrac{d\mathbf{r}}{ds}$ is a vector tangent to the curve.

In $\mathbb{R}^3$ the above relation gives

$$\frac{d\mathbf{r}}{ds} = \frac{dx}{ds}\mathbf{i} + \frac{dy}{ds}\mathbf{j} + \frac{dz}{ds}\mathbf{k} = cF_x\mathbf{i} + cF_y j + cF_z\mathbf{k}, \tag{1.76}$$

so that

$$\frac{dx}{F_x} = \frac{dy}{F_y} = \frac{dz}{F_z}, \tag{1.77}$$

represent the system of 3 differential equations whose solution gives the lines of force of $\mathbf{F}$.

Note 4
From its definition, the tangent to a line of force in every point P of the space gives the direction of the vector field evaluated in that point; it is so possible to assign an orientation to the line as that of the vector field in that point.

Exercise 1.28 Determine the lines of force of a generic central field, expressed, in $\mathbb{R}^3$, by

$$\mathbf{F} = f(x, y, z)\mathbf{e}_r = \frac{f(x, y, z)}{r}(x\mathbf{i} + y\mathbf{j} + z\mathbf{k}).$$

Solution In this case, the conditions 1.77 are written as

$$\frac{dx}{F_x} = \frac{dx}{f(x/r)} = \frac{dy}{F_y} = \frac{dy}{f(y/r)} = \frac{dz}{F_z} = \frac{dz}{f(z/r)},$$

equivalent to

$$\frac{dx}{x} = \frac{dy}{y} = \frac{dz}{z}.$$

Integrating the above equations, we have

$$\ln|y| = \ln|x| + c_1, \quad \ln|z| = \ln|x| + c_2, \quad \ln|z| = \ln|y| + c_3,$$

that is

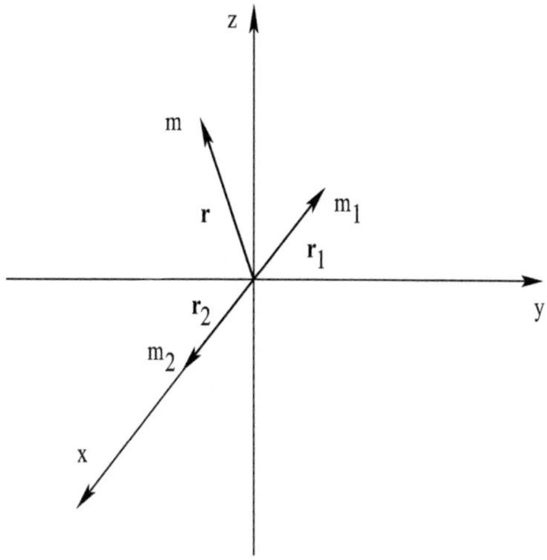

Fig. 1.4 Positions of the two particles of masses m_1 e m_2 placed on the x-axis

$$|y| = e^{c_1}|x|, \quad |z| = e^{c_2}|x|, \quad |z| = e^{c_3}|y|. \tag{1.78}$$

The first of Eqs. 1.78 corresponds, in the xy-plane, to $y = e^{c_1}x$ (1st and 4th quadrant) and $y = e^{c_1}(-x)$ (2nd and 3rd quadrant), which are straight lines passing through the origin. The same holds for the lines of force in the xz- and yz-planes.

Exercise 1.29 Determine the lines of force of the gravitational field generated by two points of masses m_1 and m_2 placed at $P_1 = (-a, 0, 0)$ and $P_2 = (a, 0, 0)$, where $a > 0$. Discuss the existence of an equilibrium in the case $m_1 = m_2$.

Solution Let $\mathbf{r}_1 = (-a, 0, 0)$ and $\mathbf{r}_2 = (a, 0, 0)$ the 2 particles position vectors; on the generic point P with radius vector $\mathbf{r} = (x, y, z)$ and unitary mass, the combined force exerted by the 2 particles is (Fig. 1.4)

$$\mathbf{F}(\mathbf{r}) = G\frac{m_1}{\|\mathbf{r}_1 - \mathbf{r}\|^3}(\mathbf{r}_1 - \mathbf{r}) + G\frac{m_2}{\|\mathbf{r}_2 - \mathbf{r}\|^3}(\mathbf{r}_2 - \mathbf{r}),$$

whose components are

$$F_x = G\frac{m_1}{\|\mathbf{r}_1 - \mathbf{r}\|^3}(-a - x) + G\frac{m_2}{\|\mathbf{r}_2 - \mathbf{r}\|^3}(a - x),$$

$$F_y = G\frac{m_1}{\|\mathbf{r}_1 - \mathbf{r}\|^3}(-y) + G\frac{m_2}{\|\mathbf{r}_2 - \mathbf{r}\|^3}(-y),$$

$$F_z = G\frac{m_1}{\|\mathbf{r}_1 - \mathbf{r}\|^3}(-z) + G\frac{m_2}{\|\mathbf{r}_2 - \mathbf{r}\|^3}(-z).$$

The conditions to determine the lines of force are

$$\frac{\mathrm{d}x}{F_x} = \frac{\mathrm{d}y}{F_y} = \frac{\mathrm{d}z}{F_z},$$

which lead to

$$\frac{\mathrm{d}x}{G\frac{m_1}{\|\mathbf{r}_1 - \mathbf{r}\|^3}(-a - x) + G\frac{m_2}{\|\mathbf{r}_2 - \mathbf{r}\|^3}(a - x)} =$$
$$= \frac{\mathrm{d}y}{G\frac{m_1}{\|\mathbf{r}_1 - \mathbf{r}\|^3}(-y) + G\frac{m_2}{\|\mathbf{r}_2 - \mathbf{r}\|^3}(-y)} =$$
$$= \frac{\mathrm{d}z}{G\frac{m_1}{\|\mathbf{r}_1 - \mathbf{r}\|^3}(-z) + G\frac{m_2}{\|\mathbf{r}_2 - \mathbf{r}\|^3}(-z)}.$$

The rightmost equality above reduces to

$$\frac{\mathrm{d}y}{y} = \frac{\mathrm{d}z}{z},$$

or, equivalently,

$$\ln|y| = \ln|z| + c_1 \iff |y| = \mathrm{e}^{c_1}|z|.$$

This means that the lines of force lie on planes through the x-axis, which means an invariance by rotation around that axis and so the lines of force are contained in surfaces of revolution around the x-axis. This is made clear by taking $\mathrm{d}y/y = \mathrm{d}z/z$ and by multiplying and dividing by y e z, so to have

$$\frac{\mathrm{d}y}{y} = \frac{\mathrm{d}z}{z} = \frac{y\mathrm{d}y}{y^2} = \frac{z\mathrm{d}z}{z^2} = \frac{1}{2}\frac{\mathrm{d}y^2}{y^2} = \frac{1}{2}\frac{\mathrm{d}z^2}{z^2} = \frac{1}{2}\frac{\mathrm{d}(y^2 + z^2)}{y^2 + z^2} = \frac{\mathrm{d}x}{F_x}.$$

The last equality is a differential relation linking x and $y^2 + z^2$, thus the solutions are ∞^1 surfaces of revolution around the x-axis.

The conclusion of the exercise is left to the reader.

1.10 The Divergence Theorem

Definition 1.5 Let $x(t)$, $y(t)$, and $z(t)$ be single-valued functions continuous in $[t_1, t_2]$ and such that

$$x(t_1) = x(t_2), \quad y(t_1) = y(t_2), \quad z(t_1) = z(t_2). \tag{1.79}$$

Under these conditions the curve in the parametric form $x = x(t)$, $y = y(t)$ and $z = z(t)$ is a *closed* curve.

Note 5
The additional condition that $x(u) = x(v)$, $y(u) = y(v)$, and $z(u) = z(v)$ holds only if $u = v$ ensures that the curve has no self-intersections: such a curve is called *simple*. Together with the previous condition, the curve is called closed and simple.

Let us suppose that the functions $x(t)$, $y(t)$ and $z(t)$ are piecewise differentiable in the interval $[t_1, t_2]$, and let give the following

Definition 1.6 A domain N of $\mathbb{R}^3$ is called *normal* if either

(*i*) N is a convex polyhedron, or
(*ii*) N is bounded by a surface S composed of a finite number of planar faces and a curved surface F, such that for some orientation of a Cartesian reference frame $O(x, y, z)$ the following hold:

(*a*) the projection $\overline{F}$ of F onto the xy-plane is bounded by a simple and closed curve consisting of a finite number of arcs, each having a continuous tangent; moreover, the projection of every face of S onto the xy-plane partitions this plane into a finite number of regions, each bounded by a simple, closed curve;
(*b*) any straight line parallel to the z-axis intersects N in a single connected segment; further, F can be represented by a Cartesian functional relation $z = f(x, y)$, where $f(x, y)$ is a single-valued function continuous with its first-order partial derivatives in the region $\overline{F}$;
(*c*) the same conditions hold for the projections onto the xz- and yz-planes.

Note 6
A sphere (solid ball) is not a normal domain because condition (ii) $b)$ does not hold globally on its curved surface F; however, it can be represented as a finite union of normal domains.

Note 7
A sub region OABC of a sphere (see Fig. 1.5) is normal as far as the central angles $\widehat{AOC}$, $\widehat{AOB}$ and $\widehat{BOC}$ are small enough.

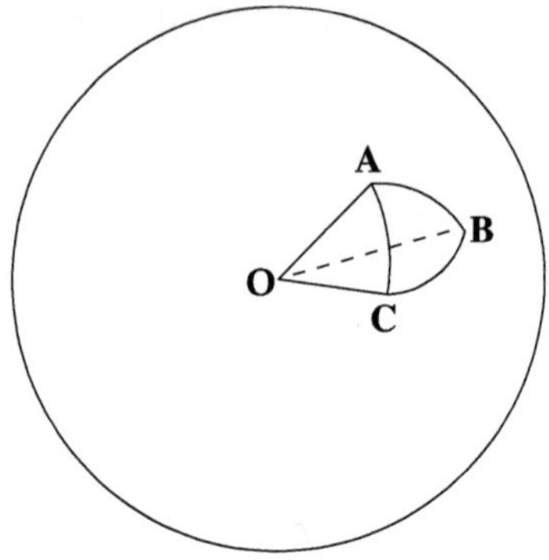

Fig. 1.5 The spherical sector O-ABC is a normal domain only when the central angles $\widehat{AOC}$, $\widehat{AOB}$ and $\widehat{BOC}$ are small enough

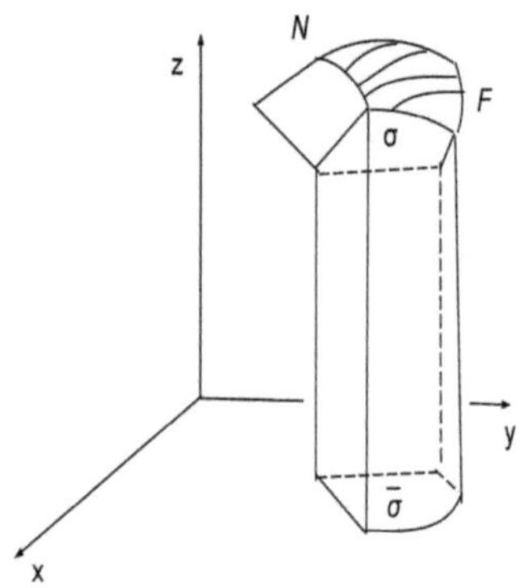

Fig. 1.6 Schematic representation of the normal domain N

Theorem 1.2 (Divergence Theorem) *Let* **A**(x,y,z) *a vector field continuous with its first order partial derivatives. If N is a normal domain, then the divergence theorem (referred also to as Ostrogradskij's theorem or, also, Gauss's theorem) holds. The thesis of the theorem is:*

$$\int_N \nabla \cdot \mathbf{A}\, \mathrm{d}V = \int_\Sigma \mathbf{A} \cdot \mathbf{n}\, \mathrm{d}\sigma, \tag{1.80}$$

where $\Sigma \equiv \partial N$ *is the surface enclosing N and* **n** *is its outward unit vector.*

Proof Let Σ the surface enclosing N, while $\overline{\sigma}$ is the projection onto the xy-plane of one, σ, of the sides of Σ, and v is the spatial region given by the intersection of N with the cylindrical surface along the sides of $\overline{\sigma}$ (Fig. 1.6). The region v is thus characterized by two surfaces $z = f_1(x, y)$ e $z = f_2(x, y)$ such that

$$f_1(x, y) \leq f_2(x, y),$$

where one of the 2 surfaces is planar.

Let us prove the theorem in the region v.

Consider, first, only the vertical component, A_z, of $\mathbf{A}$.

$$\int_v \frac{\partial A_z}{\partial z} \mathrm{d}V = \iiint_N \frac{\partial A_z}{\partial z} \mathrm{d}x\mathrm{d}y\mathrm{d}z = \iint_{\overline{\sigma}} \left[\int_{z=f_1(x,y)}^{z=f_2(x,y)} \frac{\partial A_z}{\partial z} dz \right] \mathrm{d}x\mathrm{d}y =$$

$$= \iint_{\overline{\sigma}} \Big[A_z(x, y, f_2(x, y)) - A_z(x, y, f_1(x, y))\Big]\mathrm{d}x\mathrm{d}y.$$

Indicating with $\Delta\sigma_2$ an element of the upper, curved, portion F of the surface Σ, which is represented by $z = f_2(x, y)$, and with $\overline{\Delta\sigma_2}$ the projection of $\Delta\sigma_2$ onto the xy-plane, it results

$$\overline{\Delta\sigma_2} = \int_{\overline{\Delta\sigma_2}} \mathrm{d}\overline{\sigma} = \int_{\Delta\sigma_2} \cos\alpha \, \mathrm{d}\sigma = \int_{\Delta\sigma_2} \mathbf{k} \cdot \mathbf{n}_2 \, \mathrm{d}\sigma_2,$$

where α is the angle between the outward normal, $\mathbf{n}_2$, to $\Delta\sigma_2$ and the positive z-axis.

On another side, on the lower surface (that represented by $z = f_1(x, y)$) we have

$$\overline{\Delta\sigma_1} = \int_{\overline{\Delta\sigma_1}} \mathrm{d}\overline{\sigma} = -\int_{\Delta\sigma_1} \cos\beta \, \mathrm{d}\sigma = \int_{\Delta\sigma_1} -\mathbf{k} \cdot \mathbf{n}_1 \, \mathrm{d}\sigma_1;$$

with β the angle between the outward normal, $\mathbf{n}_1$, to $\Delta\sigma_1$ and the positive z-axis; consequently

$$\iint_{\overline{\sigma}} \Big[A_z(x, y, f_2) - A_z(x, y, f_1)\Big]\mathrm{d}x\mathrm{d}y =$$

$$= \iint_{\overline{\sigma}_2} A_z(x, y, f_2)\, dxdy - \iint_{\overline{\sigma}_1} A_z(x, y, f_1)\, \mathrm{d}x\mathrm{d}y =$$

$$= \int_{\sigma_2} A_z(x, y, z)\mathbf{k} \cdot \mathbf{n}\, \mathrm{d}\sigma_2 + \int_{\sigma_1} A_z(x, y, z)\mathbf{k} \cdot \mathbf{n}\, \mathrm{d}\sigma_1 =$$

$$= \int_{\sigma_1 \cup \sigma_2 = \sigma_z} A_z\, \mathbf{k} \cdot \mathbf{n}\, \mathrm{d}\sigma = \int_{\partial v} A_z\, \mathbf{k} \cdot \mathbf{n}\, \mathrm{d}\sigma,$$

taking into account that on the lateral sides of ∂v it is, obviously, $\mathbf{k} \cdot \mathbf{n} = 0$.

Thanks to the property (ii) (c) of the normal domains, the same result above holds, also, for A_x and A_y

$$\iiint_{v} \frac{\partial A_x}{\partial x} \, \mathrm{d}x\mathrm{d}y\mathrm{d}z = \int_{\partial v} A_x \, \mathbf{i} \cdot \mathbf{n} \, \mathrm{d}\sigma,$$

$$\iiint_{v} \frac{\partial A_y}{\partial y} \, \mathrm{d}x\mathrm{d}y\mathrm{d}z = \int_{\partial v} A_y \, \mathbf{j} \cdot \mathbf{n} \, \mathrm{d}\sigma.$$

Now, due to property (ii) (a), the projection of all sides of Σ onto a coordinate plane is a finite number, n, of regions each enclosed by simple and closed curves, and so

$$\iiint_{N} \frac{\partial A_x}{\partial x} \, \mathrm{d}x\mathrm{d}y\mathrm{d}z = \sum_{k=1}^{n} \iiint_{v_k} \frac{\partial A_x}{\partial x} \, \mathrm{d}x\mathrm{d}y\mathrm{d}z = \int_{\Sigma} A_x \mathbf{i} \cdot \mathbf{n} \mathrm{d}\sigma.$$

This occurs, in perfect analogy, also for A_y e A_z, and so by summation of the integrals in the leftmost and rightmost terms member by member, and by noting that

$$A_x \, \mathbf{i} \cdot \mathbf{n} + A_y \, \mathbf{j} \cdot \mathbf{n} + A_z \, \mathbf{k} \cdot \mathbf{n} = \mathbf{A} \cdot \mathbf{n},$$

the thesis of the theorem is obtained:

$$\int_{N} \nabla \cdot \mathbf{A} \, \mathrm{d}V = \int_{\Sigma} \mathbf{A} \cdot \mathbf{n} \, \mathrm{d}\sigma.$$

Note 8

The divergence theorem can be extended to domains for which property (ii) (b) of definition 1.6 does not hold; it suffices to divide the domain in sub-regions for any of which (b) actually holds. Obviously, the outward normal to the surface of separation bounding two sub-regions has to be considered when performing the surface integral, once with a direction and once with its opposite, so that the whole contribution of such separation surface to the surface integral cancels out, leaving only the contribution of the overall bounding surface. As example, think of a spherical domain and intersect with a diametral plane. The planar cross section is the separation surface mentioned above.

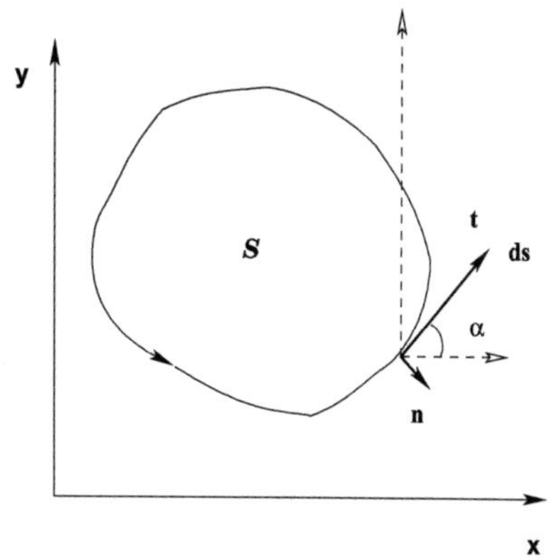

Fig. 1.7 Graphical sketch for the divergence theorem in $\mathbb{R}^2$

Theorem 1.3 (Divergence Theorem in the Plane) *Let* $\mathbf{A}$ *be a vector field, continuous with its first order partial derivatives, and* S *a planar normal domain, such that* $\bar{S} = S \cup \partial S$ *and* $C = \partial S$ *where* C *is a simple, closed planar curve. Show that in* $\mathbb{R}^2$ *the divergence theorem assumes the form (Green's theorem)*

$$\int\limits_S \left(\frac{\partial A_x}{\partial x} + \frac{\partial A_y}{\partial y} \right) \mathrm{d}\sigma = \oint\limits_C \mathbf{A} \cdot \mathbf{n}\, \mathrm{d}s = \oint\limits_C \left(A_x \mathrm{d}y - A_y \mathrm{d}x \right). \tag{1.81}$$

Proof Indicating by $\mathbf{ds}$ the line element vector counter-clockwise oriented along the contour of S, and by α the angle it forms with the direction parallel to the x-axis, from Fig. 1.7 it is clear that

$$\begin{aligned} \mathrm{d}x &= \cos\alpha\, \mathrm{d}s = \sin(\pi/2 - \alpha)\, \mathrm{d}s = -\mathbf{n} \cdot \mathbf{j}\, \mathrm{d}s = -n_y\, \mathrm{d}s, \\ \mathrm{d}y &= \sin\alpha\, \mathrm{d}s = \cos(\pi/2 - \alpha)\, \mathrm{d}s = -\mathbf{n} \cdot \mathbf{i}\, \mathrm{d}s = n_x\, \mathrm{d}s, \end{aligned}$$

by which

$$\oint\limits_C \mathbf{A} \cdot \mathbf{n}\, \mathrm{d}s = \int\limits_C \left(A_x n_x + A_y n_y \right) \mathrm{d}s = \int\limits_C \left(A_x\, \mathrm{d}y - A_y\, \mathrm{d}x \right).$$

1.10.1 The Stokes' Theorem

Let us introduce now the *Stokes' theorem.*

Definition 1.7 A surface S is said *open*, two sided (or *orientable*), when the outward normal vector is univoquely defined at every point of the surface.

Fig. 1.8 A Möbius strip (*Image credit: University College London*)

Note 9
From the above definition, a torus is an open, two sided (orientable) surface while the Möbius band (see Fig. 1.8) and the Klein bottle are not.

Theorem 1.4 (Stokes' Theorem) *Let S be any open two-sided surface bounded by a simple closed curve C, and* **A** *a vector field continuous with its first order partial derivatives in a region containing S. Under these hypotheses, it results*

$$\oint_C \mathbf{A} \cdot \mathrm{d}\mathbf{r} = \int_S \nabla \wedge \mathbf{A} \cdot \mathbf{n}\, \mathrm{d}\sigma. \tag{1.82}$$

In words, the line integral of the tangential component of **A** *along C equals the surface integral of the normal component of* $\nabla \wedge \mathbf{A}$ *on every open, two-sided surface having C as contour.*

Proof Let us prove the theorem in the hypothesis that S satisfies the same conditions of the curved surface F in the divergence theorem, i.e. it satisfies the conditions a), b) and c) of the definition 1.6 of a normal domain, and that its projections onto the coordinate planes are regions where the Green's theorem (divergence theorem in $\mathbb{R}^2$) holds. We have to show that

$$\int_S (\nabla \wedge \mathbf{A}) \cdot \mathbf{n}\, \mathrm{d}\sigma = \int_S \nabla \wedge \left(A_x \mathbf{i} + A_y \mathbf{j} + A_z \mathbf{k}\right) \cdot \mathbf{n}\, \mathrm{d}\sigma = \oint_C \mathbf{A} \cdot \mathrm{d}\mathbf{r}.$$

Since

$$\nabla \wedge A_x \mathbf{i} = \frac{\partial A_x}{\partial z}\mathbf{j} - \frac{\partial A_x}{\partial y}\mathbf{k},$$

we can write

$$\int_S (\nabla \wedge A_x \mathbf{i}) \cdot \mathbf{n}\, d\sigma = \int_S \left(\frac{\partial A_x}{\partial z} \mathbf{j} \cdot \mathbf{n} - \frac{\partial A_x}{\partial y} \mathbf{k} \cdot \mathbf{n} \right) d\sigma. \tag{1.83}$$

If the Cartesian expression for S is $z = f(x, y)$, then A_x over S is, actually, $A_x(x, y, f(x, y))$, and so

$$A_x(x, y, f(x, y)) \equiv F(x, y),$$
$$\frac{\partial F}{\partial y} = \frac{\partial A_x}{\partial y} + \frac{\partial A_x}{\partial z} \frac{\partial f}{\partial y},$$

which, by substitution of $\frac{\partial A_x}{\partial z}$ by means of the above relations, allows writing the rhs of Eq. 1.83 as

$$\int_S \left[\left(\frac{\partial F}{\partial y} - \frac{\partial A_x}{\partial y} \right) \frac{\mathbf{j} \cdot \mathbf{n}}{\frac{\partial f}{\partial y}} - \frac{\partial A_x}{\partial y} \mathbf{k} \cdot \mathbf{n} \right] d\sigma. \tag{1.84}$$

The position vector of a generic point on S is $\mathbf{r} = x\mathbf{i} + y\mathbf{j} + f(x, y)\mathbf{k}$, so that $\frac{\partial \mathbf{r}}{\partial y} = \mathbf{j} + \frac{\partial f}{\partial y}\mathbf{k}$ is a vector lying on the plane tangent, in this point, to S. As a consequence

$$\mathbf{n} \cdot \frac{\partial \mathbf{r}}{\partial y} = 0 = \mathbf{n} \cdot \mathbf{j} + \frac{\partial f}{\partial y} \mathbf{n} \cdot \mathbf{k} \Longleftrightarrow \frac{\partial f}{\partial y} = -\frac{\mathbf{n} \cdot \mathbf{j}}{\mathbf{n} \cdot \mathbf{k}};$$

which reduces the integral 1.84 to

$$-\int_S \frac{\partial F}{\partial y} \mathbf{n} \cdot \mathbf{k}\, d\sigma = -\int_{\bar{S}} \frac{\partial F}{\partial y}\, d\bar{S} = -\int\int_{\bar{S}} \frac{\partial F}{\partial y}\, dx dy,$$

where $\bar{S}$ is the projection of S onto the xy-plane (Fig. 1.9). For the divergence theorem in $\mathbb{R}^2$, it is possible to write

$$-\int\int_{\bar{S}} \frac{\partial F}{\partial y}\, dx dy = \oint_{\partial \bar{S}} F\, dx,$$

where $\partial \bar{S}$ is the contour of $\bar{S}$, which is a simple closed curve, Γ. Finally, we have

$$\int_S (\nabla \wedge A_x \mathbf{i}) \cdot \mathbf{n}.\, d\sigma = \oint_\Gamma F\, dx.$$

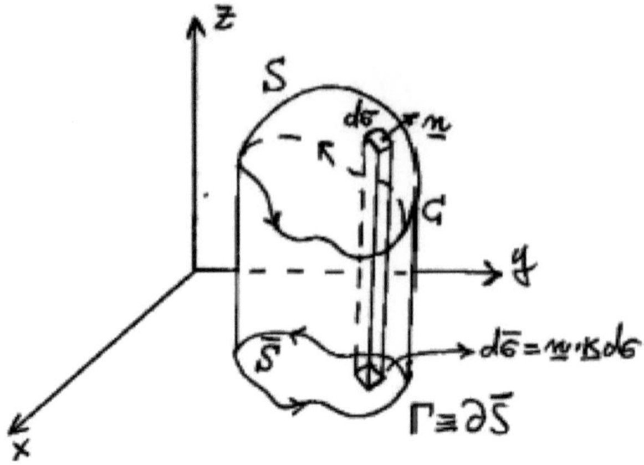

Fig. 1.9 Schematic representation of the surface S and of its projection $\bar{S}$ onto the xy-plane

Noting that $F(x, y) = A_x(x, y, f(x, y))$, we have that F over Γ is equal to A_x over C and, due to that dx is the same on C and on Γ, it results

$$\int_S (\nabla \wedge A_x \mathbf{i}) \cdot \mathbf{n} \, \mathrm{d}\sigma = \oint_C A_x \, \mathrm{d}x. \tag{1.85}$$

Similarly, on the xz- and yz-planes it is

$$\int_S (\nabla \wedge A_y \mathbf{j}) \cdot \mathbf{n} \, \mathrm{d}\sigma = \oint_C A_y \, \mathrm{d}y, \tag{1.86}$$

$$\int_S (\nabla \wedge A_z \mathbf{k}) \cdot \mathbf{n} \, \mathrm{d}\sigma = \oint_C A_z \, \mathrm{d}z. \tag{1.87}$$

The thesis of the theorem is so obtained by direct summation of the lhs and rhs of (1.85), (1.86), and (1.87), getting

$$\int_S (\nabla \wedge \mathbf{A}) \cdot \mathbf{n} \, \mathrm{d}\sigma = \oint_C \mathbf{A} \cdot \mathrm{d}\mathbf{r}.$$

Corollary 1.2 *A direct consequence of the Stokes' Theorem is*

$$\nabla \wedge \mathbf{A} = 0 \quad \Longleftrightarrow \quad \oint_C \mathbf{A} \cdot \mathrm{d}\mathbf{r} = 0, \tag{1.88}$$

where C is an arbitrary simple closed curve.

The relation 1.82 implies

$$(i) \qquad \nabla \wedge \mathbf{A} = 0 \implies \oint_C \mathbf{A} \cdot \mathrm{d}\mathbf{r} = 0,$$

and, by the continuity of both **A** *and its first order partial derivatives, it results also*

$$(ii) \qquad \oint_C \mathbf{A} \cdot \mathrm{d}\mathbf{r} = 0 \implies \nabla \wedge \mathbf{A} = 0.$$

Relation (ii) *is obtained by considering that, if there exists a point* P *where* $\nabla \wedge \mathbf{A} \neq 0$*, then, by continuity, it should exist a whole neighbourhood* V *of* P *where* $\nabla \wedge \mathbf{A} \neq 0$ *and so, taking a surface* $S \subset V$ *and its outward normal unit vector* **n***, it would be* $\nabla \wedge \mathbf{A} \cdot \mathbf{n} \neq 0$ *on the surface, and thus*

$$\oint_{C_S} \mathbf{A} \cdot \mathrm{d}\mathbf{r} = \int_S \nabla \wedge \mathbf{A} \cdot \mathbf{n} \, \mathrm{d}\sigma \neq 0$$

where C_S *is the curve bounding* S*, against the hypothesis.*

Exercise 1.30 Given, in $\mathbb{R}^2$, the vector field

$$\mathbf{F} = \frac{y}{r^2}\mathbf{i} - \frac{x}{r^2}\mathbf{j},$$

show that for it the Stokes' theorem does not hold over the all xy-plane.

Solution The curl of **F** in two dimensions is

$$\nabla \wedge \mathbf{F} = \mathbf{k}\left(\frac{\partial F_y}{\partial x} - \frac{\partial F_x}{\partial y}\right),$$

which vanishes everywhere (but the origin where it is not defined) because

$$\frac{\partial F_x}{\partial y} = \frac{r^2 - 2yr\frac{y}{r}}{r^4} = \frac{x^2 - y^2}{r^4},$$

$$\frac{\partial F_y}{\partial x} = \frac{-r^2 + 2xr\frac{x}{r}}{r^4} = \frac{x^2 - y^2}{r^4}.$$

If $\nabla \wedge \mathbf{F} = 0$ everywhere but the origin, obviously

$$\int_S (\nabla \wedge \mathbf{F}) \cdot \mathbf{n} \, \mathrm{d}\sigma = 0,$$

for any planar region S *not* containing the origin.

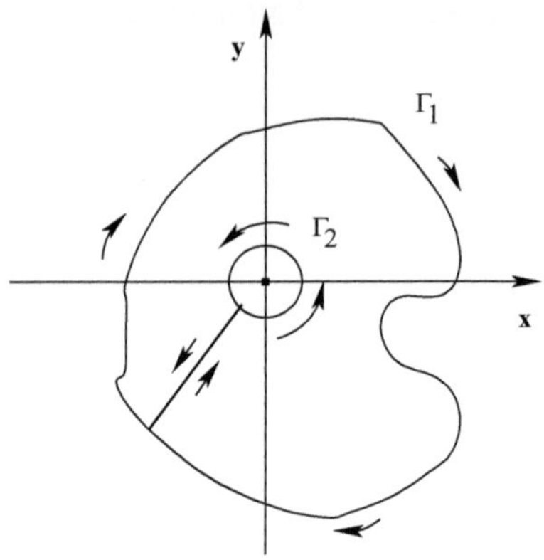

Fig. 1.10 Path apt to exclude the singularity of $\mathbf{F}(x, y)$ in the origin

Note that, if we perform the line integral $\oint \mathbf{F} \cdot \mathbf{dr}$ along a circumference C centred at the coordinate origin by using polar coordinates (r, θ), we have

$$\oint_C \mathbf{F} \cdot \mathbf{dr} = \int_0^{2\pi} \frac{1}{r^2}(-r^2 \cos^2 \theta - r^2 \sin^2 \theta)\, d\theta = -\int_0^{2\pi} d\theta = -2\pi.$$

This result does not contradict Stokes' theorem, because one of the fundamental assumptions of the theorem is violated: the vector field must be continuously differentiable on an open surface containing the curve and its interior. Here, the origin lies inside the curve, but $\mathbf{F}$ is undefined there, so the theorem does not apply.

To better understand the role of the singularity, consider a path $\Gamma = \Gamma_1 \cup \Gamma_2$ forming a closed loop that encloses an annular region and excludes the origin, as illustrated in Fig. 1.10. In this punctured domain, the field is regular, and Stokes' theorem holds. Since the curl of $\mathbf{F}$ is zero throughout the annulus, we have:

$$\oint_{\Gamma_1} \mathbf{F} \cdot \mathbf{dr} - \oint_{\Gamma_2} \mathbf{F} \cdot \mathbf{dr} = \int_{\text{annulus}} (\nabla \wedge \mathbf{F}) \cdot \mathbf{n}\, d\sigma = 0,$$

which implies:

$$\oint_{\Gamma_1} \mathbf{F} \cdot \mathbf{dr} = \oint_{\Gamma_2} \mathbf{F} \cdot \mathbf{dr}.$$

This shows that the nonzero line integral is not due to local behaviour of the field, but rather to the *topological obstruction* introduced by the singularity at the origin. The integral captures a kind of *circulation "residue"*, which is a global feature of the domain. While the singularity can be avoided locally by choosing appropriate integration paths, its effect cannot be removed globally—it reveals that the field is not conservative, despite having zero curl almost everywhere.

In $\mathbb{R}^2 \setminus (0, 0)$ the potential of $\mathbf{F}$ is

$$U(x, y) = \arctan\left(\frac{x}{y}\right) + c,$$

as it is easily verified.

Exercise 1.31 If $S = S_1 \cup S_2$ is a closed surface union of two surfaces for which the Stokes' theorem holds, show that

$$\int_S \nabla \wedge \mathbf{A} \cdot \mathbf{n}\, d\sigma = 0.$$

Solution Left to the reader.

Exercise 1.32 Let $S_R(0)$ a sphere of radius R centred at the coordinate origin, and let $\mathbf{A} = a(x, y, z)\, \mathbf{e}_r$ a continuous vector field; calculate

$$\int_{S_R(0)} \nabla \cdot \mathbf{A}\, dV.$$

Solution By the divergence theorem, we have

$$\int_{S_R(0)} \nabla \cdot \mathbf{A}\, dV = \int_{\partial S_R(0)} \mathbf{A} \cdot \mathbf{n}\, d\sigma,$$

that, in spherical polar coordinates ($d\sigma = r^2 \sin\varphi\, d\theta d\varphi$), reads as

$$\int_{\partial S_R(0)} \mathbf{A} \cdot \mathbf{n}\, d\sigma = R^2 \int_0^{\pi}\int_0^{2\pi} a(x, y, z) \sin\varphi\, d\theta d\varphi.$$

In the simplifying assumption that $a(x, y, z) = a(r)$, it results

$$R^2 \int_0^{\pi}\int_0^{2\pi} a(R) \sin\varphi\, d\theta d\varphi = 4\pi R^2 a(R).$$

Another way to get the result is through the following development

$$\int_{S_R(0)} \nabla \cdot \mathbf{A}\, \mathrm{d}V = \int_{S_R(0)} (a\nabla \cdot \mathbf{e}_r + \mathbf{e}_r \cdot \nabla a)\ \mathrm{d}V = \int_0^R \left(a\frac{2}{r} + \frac{da}{dr}\right) 4\pi r^2\, \mathrm{d}r =$$

$$= \int_0^R \frac{1}{r^2}\frac{\mathrm{d}}{\mathrm{d}r}(r^2 a)\, 4\pi r^2\, \mathrm{d}r = 4\pi \int_0^R \mathrm{d}(r^2 a) = 4\pi R^2 a(R),$$

which vanishes only if $a(R) = 0$.

Exercise 1.33 Same as the previous exercise, in the case $a(R, \theta, \varphi) = -a(R, \theta, \pi - \varphi)$, i.e. for a vector field which is antisymmetric respect to the xy-plane.

Solution

$$\int_N \nabla \cdot \mathbf{A}\, \mathrm{d}V = R^2 \int_0^{\pi}\int_0^{2\pi} a(R, \theta, \varphi) \sin\varphi\, \mathrm{d}\theta \mathrm{d}\varphi =$$

$$= R^2 \int_0^{2\pi} \left[\int_0^{\frac{\pi}{2}} a(R, \theta, \varphi) \sin\varphi\, \mathrm{d}\varphi + \int_{\frac{\pi}{2}}^{\pi} a(R, \theta, \varphi) \sin\varphi\, \mathrm{d}\varphi \right] \mathrm{d}\theta =$$

$$= R^2 \int_0^{2\pi} \left[\int_0^{\frac{\pi}{2}} a(R, \theta, \varphi) \sin\varphi\, \mathrm{d}\varphi - \int_{\frac{\pi}{2}}^{\pi} a(R, \theta, \pi - \varphi) \sin\varphi\, \mathrm{d}\varphi \right] \mathrm{d}\theta =$$

$$= 2R^2 \int_0^{2\pi} \left[\int_0^{\frac{\pi}{2}} a(R, \theta, \varphi) \sin\varphi\, \mathrm{d}\varphi \right] \mathrm{d}\theta.$$

Exercise 1.34 Show that the radius, R, of a sphere centred in the origin can be expressed as

$$R = \left(\frac{1}{4\pi} \int_{\partial S_R(0)} \mathbf{r} \cdot \mathbf{n}\, \mathrm{d}\sigma \right)^{1/3},$$

where $\partial S_R(0)$ is the sphere surface.

Solution By the divergence theorem, it is

$$\int_{\partial S_R(0)} \mathbf{r}\cdot\mathbf{n}\,\mathrm{d}\sigma = \int_{S_R(0)} \nabla\cdot\mathbf{r}\,\mathrm{d}V = 3V_R = 4\pi R^3.$$

Exercise 1.35 Prove that the vector field

$$\mathbf{F} = (2xy+3)\,\mathbf{i} + (x^2-4z)\,\mathbf{j} - 4y\,\mathbf{k}$$

is conservative and find its potential function.

Solution The existence of the potential is guaranteed by

$$\nabla\wedge\mathbf{F} = \left(\frac{\partial F_z}{\partial y} - \frac{\partial F_y}{\partial z}\right)\mathbf{i} + \left(\frac{\partial F_z}{\partial x} - \frac{\partial F_x}{\partial z}\right)\mathbf{j} + \left(\frac{\partial F_y}{\partial x} - \frac{\partial F_x}{\partial y}\right)\mathbf{k} =$$
$$= (-4+4)\,\mathbf{i} + 0\cdot\mathbf{j} + (2x-2x)\mathbf{k} = 0.$$

We are looking for the potential, i.e. a function $U(x, y, z)$ such that

$$\frac{\partial U}{\partial x} = F_x, \qquad \frac{\partial U}{\partial y} = F_y, \qquad \frac{\partial U}{\partial z} = F_z.$$

Consequently, given the above expression of **F**, a first integration over x gives

$$U(x, y, z) = \int F_x\,\mathrm{d}x + f(y, z) = x^2y + 3x + f(y, z), \tag{1.89}$$

where $f(y, z)$ is an arbitrary function of y and z only, to be determined by derivation of the above function U with respect to y

$$\frac{\partial U}{\partial y} = x^2 + \frac{\partial f}{\partial y} = F_y = x^2 - 4z,$$

which, integrating with respect to y, gives

$$f(y, z) = -\int 4z\,\mathrm{d}y = -4zy + g(z).$$

Substituting in Eq. 1.89 we have

$$U(x, y, z) = x^2y + 3x - 4zy + g(z)$$

whose differentiation with respect to z leads to

$$\frac{\partial U}{\partial z} = -4y + \frac{\mathrm{d}g}{\mathrm{d}z} = F_z = -4y \quad\Longrightarrow\quad \frac{\mathrm{d}g}{\mathrm{d}z} = 0 \quad\Longrightarrow\quad g(z) = c,$$

which, eventually, leads to

$$U(x, y, z) = x^2 y + 3x - 4zy + c.$$

1.11 Velocity Fields

If the fluid is considered as a set of N "particles", in line of principle the perfect description of the fluid motion is through the knowledge at every time t of all the $i = 1, 2, .., N$ particle position vectors $\mathbf{r}_i(t)$, that could be determined by the knowledge of the total forces acting on every particle followed by the integration of their Newtonian equations of motion. This would involve an overwhelming number of evaluations of $O(N^2)$ pair interactions. Let us think about a one-litre bottle of water as constituted by N H_2O molecules considered as fluid elementary particles. The number of water molecules in a litre of water at room temperature is $N \simeq 3.34 \times 10^{25}$, so that the number of force interactions to compute is $N(N-1)/2 \sim 5.58 \times 10^{50}$. This means that a perfectly deterministic particle-by-particle motion calculation is absolutely unviable. For this reason, some approximation is needed.

There are two ways to approach the study of the motion of a fluid: *Eulerian* or *Lagrangian*.

The Eulerian specification (or view) corresponds to a *field* view, and the most relevant field is the velocity one.

On the other side, the Lagrangian specification of the fluid flow is a way of looking at fluid motion where the "observer" follows an individual fluid *parcel* (a representative, and small, piece of fluid) as it moves through space and time. Plotting the position of an individual parcel through time yields the pathline of the parcel. Mathematically speaking, supposedly known the position of a given ith fluid particle at a given time t_0, indicated by $\mathbf{r}_{i0} = (x_{i0}, y_{i0}, z_{i0})$, the Lagrangian description corresponds to the knowledge, at any time t, of the vector function $\mathbf{r}_i = \mathbf{r}_i(\mathbf{r}_{i0}; t)$, i.e. of the 3 scalar functions

$$\begin{cases} x_i = x_i(x_{i0}, y_{i0}, z_{i0}; t), \\ y_i = y_i(x_{i0}, y_{i0}, z_{i0}; t), \\ z_i = z_i(x_{i0}, y_{i0}, z_{i0}; t), \end{cases} \tag{1.90}$$

supposed continuous for every t (if $t \to t_0$, then $x \to x_0$, $y \to y_0$, $z \to z_0$).

The time derivatives of the functions in (1.90)

$$\begin{cases} \dot{x}_i = \dot{x}_i(x_{i0}, y_{i0}, z_{i0}; t), \\ \dot{y}_i = \dot{y}_i(x_{i0}, y_{i0}, z_{i0}; t), \\ \dot{z}_i = \dot{z}_i(x_{i0}, y_{i0}, z_{i0}; t), \end{cases} \tag{1.91}$$

provide the velocity of every particle at any time as a function of its initial position (that at time t_0). The mathematical procedure of elimination of x_{i0}, y_{i0}, z_{i0} in the rhs of Eqs. 1.91, which is done (if possible) by inversion in Eqs. 1.90, allows passing from Lagrangian to Eulerian specification, which corresponds to the identification of the velocity field (fluid *flow*) $\mathbf{v} = v_x\mathbf{i} + v_y\mathbf{j} + v_z\mathbf{k}$ at any time t and in any point (x, y, z) of the 3-dimensional space[6]

$$\dot{x} = \dot{x}(x, y, z; t) = v_x, \qquad \dot{y} = \dot{y}(x, y, z; t) = v_y, \qquad \dot{z} = \dot{z}(x, y, z; t) = v_z. \tag{1.92}$$

The velocity flow may depend explicitly on t. In that case, the flow lines[7] also depend on time and constitute, so, a set of ∞^1 curves defined by the usual (see (1.77)) differential conditions

$$\frac{\mathrm{d}x}{v_x(x, y, z; t)} = \frac{\mathrm{d}y}{v_y(x, y, z; t)} = \frac{\mathrm{d}z}{v_z(x, y, z; t)}. \tag{1.93}$$

If the velocity field is independent of t (*stationary* flux), the curves satisfying Eqs. 1.93 are the actual trajectories of fluid motion.

Exercise 1.36 Given, in a Lagrangian fluid view

$$\begin{cases} \mathbf{r}(\mathbf{r}_0; t) = x_0 \mathrm{e}^t\,\mathbf{i} + y_0 \mathrm{e}^{-t}\,\mathbf{j} + z_0 \mathrm{e}^t\,\mathbf{k}, \\ \mathbf{r}(0) = \mathbf{r}_0, \end{cases}$$

or, in scalar form

$$x = x_0 \mathrm{e}^t, \qquad y = y_0 \mathrm{e}^{-t}, \qquad z = z_0,$$

it is asked to obtain its Eulerian description.

Solution By time differentiation of the above expressions for $x(t)$, $y(t)$ and $z(t)$:

$$\dot{x} = x_0 \mathrm{e}^t, \qquad \dot{y} = -y_0 \mathrm{e}^{-t}, \qquad \dot{z} = 0, \tag{1.94}$$

thus the Eulerian form, that is obtained by elimination of x_0 and y_0 in the above relations, is

$$\dot{x} = x = v_x, \qquad \dot{y} = -y = v_y, \qquad \dot{z} = 0 = v_z.$$

This is a planar stationary flow, whose flow lines (which are trajectories of motion due to independence of t) are

$$\frac{\mathrm{d}x}{x} = \frac{\mathrm{d}y}{-y} \quad \Longrightarrow \quad \ln|x| = -\ln|y| + c,$$

[6] Dropping, for simplicity, the index i.

[7] The flow lines are the velocity field lines.

or, equivalently

$$|x||y| = \mathrm{e}^c = C, \quad \text{hyperbola.}$$

The above Cartesian expression of trajectories obviously coincides with that obtained by a straight elimination of the t parameter in Eqs. 1.94, leading to

$$x = x_0\mathrm{e}^t \text{ and } y = y_0\mathrm{e}^{-t}, \text{ so that } xy = x_0y_0\mathrm{e}^t\mathrm{e}^{-t} = x_0y_0.$$

Exercise 1.37 Given the (Eulerian) velocity field

$$\mathbf{v}(\mathbf{r};\, t) = -\cos t\,\mathbf{i} + \sin t\,\mathbf{j},$$

determine fluid trajectories and streamlines.[8]

Solution Left to the reader.

Exercise 1.38 Study the properties of the fluid motion represented by the Lagrangian expression

$$\mathbf{r}(\mathbf{r}_0;\, t) = \left(\frac{x_0 + y_0}{2}\mathrm{e}^t + \frac{x_0 - y_0}{2}\mathrm{e}^{-t}\right)\mathbf{i} + \left(\frac{x_0 + y_0}{2}\mathrm{e}^t - \frac{x_0 - y_0}{2}\mathrm{e}^{-t}\right)\mathbf{j} + z_0\,\mathbf{k}.$$

Solution Left to the reader.

1.12 Meaning of the Divergence of a Vector Field

A fluid in motion can expand or contract: a fluid element that at t_0 occupies a portion of space C_0, at time t will extend over a portion of space C such that, in general, $vol(C_0) \neq vol(C)$. In such case we speak of a *compressible* fluid. Indicating by $\rho(x, y, z; t)$ the fluid mass density in a point (x, y, z) at time t, the mass continuity equation in Lagrangian form is

$$\frac{\mathrm{D}\rho}{\mathrm{D}t} = -\rho\nabla\cdot\mathbf{v}, \tag{1.95}$$

where $\mathrm{D}/\mathrm{D}t$ represents the Lagrangian time derivative (that evaluated along the flow), and defining the specific volume $V^* = 1/\rho$ as that occupied by the unitary fluid mass, we clearly have that in the incompressible case $V^* = const.$, and thus

$$\frac{\mathrm{D}V^*}{\mathrm{D}t} = -\frac{1}{\rho^2}\frac{\mathrm{D}\rho}{\mathrm{D}t} = 0. \tag{1.96}$$

[8] A streamline is everywhere tangent to the velocity field. When the flow is steady, streamlines coincide with fluid particle trajectories.

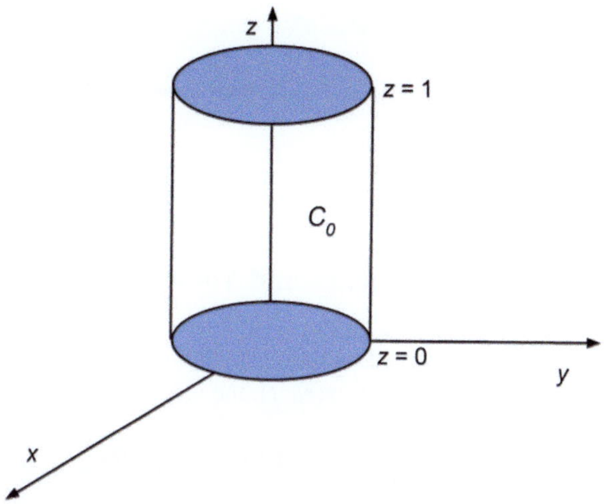

Fig. 1.11 The cylindrical domain C_0 of Exercise 1.39

So, from Eqs. 1.95 and 1.96, an incompressible flow is *solenoidal*, i.e. $\nabla \cdot \mathbf{v} = 0$.

Exercise 1.39 Let us consider the Lagrangian flow defined by

$$\mathbf{r}(\mathbf{r}_0; t) = x_0 \mathrm{e}^t \mathbf{i} + y_0 \mathrm{e}^{-t} \mathbf{j} + \mathbf{k}. \tag{1.97}$$

If at $t = t_0$ the domain C_0 is given by the cylinder (Fig. 1.11)

$$0 \leq z \leq 1, \quad x_0^2 + y_0^2 = a^2,$$

show that at a generic t the domain C_t results defined by

$$0 \leq z \leq 1, \qquad \text{and} \begin{cases} x = x_0 \mathrm{e}^t, & (x \text{ expansion}) \\ y = y_0 \mathrm{e}^{-t}, & (y \text{ compression}). \end{cases}$$

Solution From Eq. 1.97 we get

$$\frac{x^2}{\mathrm{e}^{2t}} + \frac{y^2}{\mathrm{e}^{-2t}} = x_0^2 + y_0^2 = a^2,$$

that is

$$\left(\frac{x}{a\mathrm{e}^t}\right)^2 + \left(\frac{y}{a\mathrm{e}^{-t}}\right)^2 = 1,$$

which is the equation of the ellipse of semiaxes $|a|\mathrm{e}^t$ (x expansion) and $|a|\mathrm{e}^{-t}$ (y contraction). Note the invariance of the domain volume in time

$$vol(C_t) = \pi |a| e^t |a| e^{-t} = \pi a^2 = vol(C_0),$$

in agreement with the fluid incompressibility, guaranteed by that $\nabla \cdot \mathbf{v} = 0$ (indeed, $\mathbf{v} = x\,\mathbf{i} - y\,\mathbf{j}$).

Exercise 1.40 Suggest examples of flows for which $vol(C_0) \neq vol(C)$.

1.13 Link Between Divergence and Volume Variation

If C_t is the space domain occupied, at time t, by a fluid which at time t_0 occupied C_0, in the Lagrangian framework $\mathbf{r} = \mathbf{r}(\mathbf{r}_0; t)$, the volume of C_t is expressed as

$$V(t) = \iiint\limits_{C_t} \mathrm{d}^3\mathbf{r} = \iiint\limits_{C_0} J(\mathbf{r}_0, \mathbf{r})\, \mathrm{d}^3\mathbf{r}_0, \tag{1.98}$$

where $J(\mathbf{r}_0, \mathbf{r})$ is the Jacobian determinant of the coordinate transformation from $\mathbf{r}_0$ to $\mathbf{r}$

$$J(\mathbf{r}_0, \mathbf{r}) \equiv \det\left(\frac{\partial \mathbf{r}}{\partial \mathbf{r}_0}\right) = \begin{vmatrix} \dfrac{\partial x}{\partial x_0} & \dfrac{\partial x}{\partial y_0} & \dfrac{\partial x}{\partial z_0} \\ \dfrac{\partial y}{\partial x_0} & \dfrac{\partial y}{\partial y_0} & \dfrac{\partial y}{\partial z_0} \\ \dfrac{\partial z}{\partial x_0} & \dfrac{\partial z}{\partial y_0} & \dfrac{\partial z}{\partial z_0} \end{vmatrix}. \tag{1.99}$$

If $J(\mathbf{r}_0, \mathbf{r})$ is time-differentiable with continuity, then

$$\dot{V}(t) = \iiint\limits_{C_0} \dot{J}(\mathbf{r}_0, \mathbf{r})\, \mathrm{d}^3\mathbf{r}_0. \tag{1.100}$$

The integrand in the integral above is

$$\dot{J}(\mathbf{r}_0, \mathbf{r}) = \begin{vmatrix} \dfrac{\partial^2 x}{\partial t \partial x_0} & \dfrac{\partial^2 x}{\partial t \partial y_0} & \dfrac{\partial^2 x}{\partial t \partial z_0} \\ \dfrac{\partial y}{\partial x_0} & \dfrac{\partial y}{\partial y_0} & \dfrac{\partial y}{\partial z_0} \\ \dfrac{\partial z}{\partial x_0} & \dfrac{\partial z}{\partial y_0} & \dfrac{\partial z}{\partial z_0} \end{vmatrix} + \begin{vmatrix} \dfrac{\partial x}{\partial x_0} & \dfrac{\partial x}{\partial y_0} & \dfrac{\partial x}{\partial z_0} \\ \dfrac{\partial^2 y}{\partial t \partial x_0} & \dfrac{\partial^2 y}{\partial t \partial y_0} & \dfrac{\partial^2 y}{\partial t \partial z_0} \\ \dfrac{\partial z}{\partial x_0} & \dfrac{\partial z}{\partial y_0} & \dfrac{\partial z}{\partial z_0} \end{vmatrix} + \begin{vmatrix} \dfrac{\partial x}{\partial x_0} & \dfrac{\partial x}{\partial y_0} & \dfrac{\partial x}{\partial z_0} \\ \dfrac{\partial y}{\partial x_0} & \dfrac{\partial y}{\partial y_0} & \dfrac{\partial y}{\partial z_0} \\ \dfrac{\partial^2 z}{\partial t \partial x_0} & \dfrac{\partial^2 z}{\partial t \partial y_0} & \dfrac{\partial^2 z}{\partial t \partial z_0} \end{vmatrix}. \tag{1.101}$$

Under the assumption that the first and second partial derivatives are continuous, the evaluation of $\dot{J}(\mathbf{r}_0, \mathbf{r})$ at time t_0 gives

$$\dot{J}(\mathbf{r}_0, \mathbf{r}) = \begin{vmatrix} \frac{\partial}{\partial x_0}\left(\frac{\partial x}{\partial t}\right) & \frac{\partial}{\partial y_0}\left(\frac{\partial x}{\partial t}\right) & \frac{\partial}{\partial z_0}\left(\frac{\partial x}{\partial t}\right) \\ 0 & 1 & 0 \\ 0 & 0 & 1 \end{vmatrix}$$
$$+ \begin{vmatrix} 1 & 0 & 0 \\ \frac{\partial}{\partial x_0}\left(\frac{\partial y}{\partial t}\right) & \frac{\partial}{\partial y_0}\left(\frac{\partial y}{\partial t}\right) & \frac{\partial}{\partial z_0}\left(\frac{\partial y}{\partial t}\right) \\ 0 & 0 & 1 \end{vmatrix} + \tag{1.102}$$
$$+ \begin{vmatrix} 1 & 0 & 0 \\ 0 & 1 & 0 \\ \frac{\partial}{\partial x_0}\left(\frac{\partial z}{\partial t}\right) & \frac{\partial}{\partial y_0}\left(\frac{\partial z}{\partial t}\right) & \frac{\partial}{\partial z_0}\left(\frac{\partial z}{\partial t}\right) \end{vmatrix},$$

more compactly written as

$$\dot{J}(\mathbf{r}_0, \mathbf{r}) = \begin{vmatrix} \frac{\partial v_x}{\partial x_0} & \frac{\partial v_x}{\partial y_0} & \frac{\partial v_x}{\partial z_0} \\ 0 & 1 & 0 \\ 0 & 0 & 1 \end{vmatrix} + \begin{vmatrix} 1 & 0 & 0 \\ \frac{\partial v_y}{\partial x_0} & \frac{\partial v_y}{\partial y_0} & \frac{\partial v_y}{\partial z_0} \\ 0 & 0 & 1 \end{vmatrix} + \begin{vmatrix} 1 & 0 & 0 \\ 0 & 1 & 0 \\ \frac{\partial v_z}{\partial x_0} & \frac{\partial v_z}{\partial y_0} & \frac{\partial v_z}{\partial z_0} \end{vmatrix} = \tag{1.103}$$
$$= \left(\frac{\partial v_x}{\partial x_0} + \frac{\partial v_y}{\partial y_0} + \frac{\partial v_z}{\partial z_0}\right)_{t=t_0} = (\nabla \cdot \mathbf{v})_{t=t_0}.$$

Given the arbitrariness of t_0, we have for every t

$$\dot{V}(t) = \iiint\limits_{C_t} \nabla \cdot \mathbf{v}\, \mathrm{d}^3\mathbf{r}. \tag{1.104}$$

As a consequence, if $\nabla \cdot \mathbf{v} = 0$ everywhere in space, then

$$\dot{V}(t) = 0 \iff V(t) = constant. \tag{1.105}$$

Vice versa, if $\dot{V}(t) = 0 \quad \forall t$, then the continuity of $\mathbf{v}$ and of its first derivatives imply $\nabla \cdot \mathbf{v} = 0$: the fluid is said to be *incompressible*. Applying the mean theorem for integrals, we have

$$\frac{\dot{V}}{V} = \langle \nabla \cdot \mathbf{v} \rangle, \qquad \lim_{C \to (x,y,z)} \frac{\dot{V}}{V} = \nabla \cdot \mathbf{v},$$

where the brackets represent an average over the domain. Clearly, $\nabla \cdot \mathbf{v}$ is the fractional rate of expansion ($\nabla \cdot \mathbf{v} > 0$) or compression ($\nabla \cdot \mathbf{v} < 0$) of the fluid around the generic point (x, y, z).

Exercise 1.41 Show that the volume of a fluid domain C is expressible by the formula

$$V(C) = \frac{1}{3}\int_{\partial C} \mathbf{r}\cdot\mathbf{n}\,d\sigma.$$

Solution Defining the vector field $\mathbf{F} = \mathbf{r}/3$, its divergence is $\nabla\cdot(\mathbf{r}/3) = 1$, so that, applying the divergence theorem, we have

$$\int_C \nabla\cdot\mathbf{F}dV = \int_C dV = V(C) = \int_{\partial C} \mathbf{F}\cdot\mathbf{n}\,d\sigma = \frac{1}{3}\int_{\partial C} \mathbf{r}\cdot\mathbf{n}\,d\sigma.$$

1.14 Flux Tubes

The original description of a flux tube is due to J. C. Maxwell ("On Physical Lines of Force", *Philosophical Magazine and Journal of Science*, vol. 4, 1861) who, referring to a fluid velocity field, said: "If upon any surface which cuts the lines of fluid motion we draw a closed curve, and if from every point of this curve we draw lines of motion, these lines of motion will generate a tubular surface which we may call a tube of fluid motion".

Although lacking of mathematical rigour, this can be indeed considered a good operational definition of a generic vector field tube.

Let us consider a region of space T bounded by a flux tube and by two cross sections S_1 and S_2 orthogonal to the tube, as in Fig. 1.12.

If the generic vector field $\mathbf{F}$ is solenoidal (like the velocity field of an incompressible fluid) then

$$\int_T \nabla\cdot\mathbf{F}\,dV = \int_{\partial T} \mathbf{F}\cdot\mathbf{n}\,d\sigma = \int_{S_1\cup S_2} \mathbf{F}\cdot\mathbf{n}\,d\sigma + \int_L \mathbf{F}\cdot\mathbf{n}\,d\sigma = 0, \tag{1.106}$$

where ∂T is the union of S_1, S_2 and of the lateral surface, L, of the tube (see Fig. 1.12). Clearly, the surface integral over the lateral surface is zero because there

Fig. 1.12 A flux tube

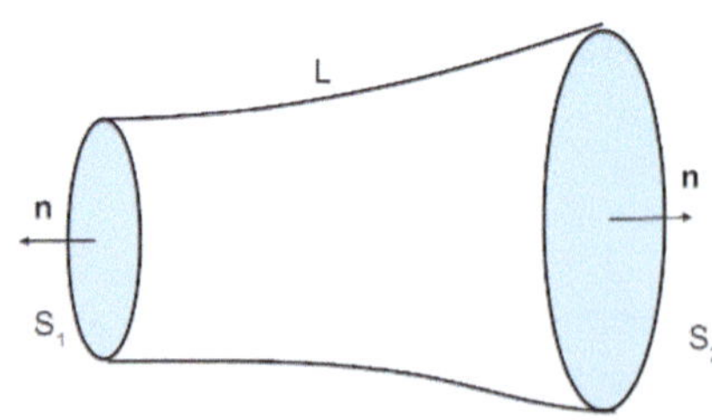

$\mathbf{F} \cdot \mathbf{n} = 0$, so that Eq, 1.106 reduces to

$$\int\limits_{S_1 \cup S_2} \mathbf{F} \cdot \mathbf{n} \, d\sigma = 0, \tag{1.107}$$

which implies

$$\int\limits_{S_1} \mathbf{F} \cdot \mathbf{n} \, d\sigma = -\int\limits_{S_2} \mathbf{F} \cdot \mathbf{n} \, d\sigma. \tag{1.108}$$

A straight application of the integral mean theorem leads to

$$\langle F_1 \rangle area(S_1) = \langle F_2 \rangle area(S_2), \tag{1.109}$$

where F_1 and F_2 are the norms of the $\mathbf{F}$ vector field outward normal components over the two cross sectional surfaces. This, obviously, leads to the scaling

$$\frac{\langle F_1 \rangle}{\langle F_2 \rangle} = \frac{area(S_2)}{area(S_1)}, \tag{1.110}$$

that is: the intensity of a solenoidal vector field increases (the field line density increases) where the flux tube shrinks while its intensity decreases (as the field lines density) across larger cross sections of the flux tube. The ratio is inversely proportional to the area of the orthogonal cross sections. The formula 1.109 is often referred to as Gauss' law.

1.15 Further Readings

This chapter contains all the concepts necessary to understand the remainder of the book. Readers who wish to deepen their knowledge of the theory of functions of many variables and vector analysis can refer to many excellent introductory textbooks like [3] and [4]. Relevant aspects of fluid dynamics are covered by Landau and Lifshits [2] and Capuzzo Dolcetta [1].

References

1. Capuzzo Dolcetta, R.A.: Physics of Fluids. Unitext for Physics, Springer Nature, Cham (2024)
2. Landau, L.D., Lifshits, E.M.: Fluid Dynamics, Course of Theoretical Physics, vol. 6, 2nd edn. Pergamon Press Oxford (1989)
3. Spiegel, M.R.: Advanced Mathematics. Schaum's Outline Series, McGraw Hill, New York (2010)
4. Wrede, R.C.: Introduction to Vector and Tensor Analysis. Dover Publications Inc., New York (1972)

Chapter 2
Newtonian Gravitational Interaction

This chapter presents the classical Newtonian approach to gravitation, both in the framework of discrete N-body systems and in that of continuous matter distributions.

We begin with the particular case of the one-body problem, and treat the motion of a point mass in a resisting medium. We then introduce the general N-body problem in the Newtonian context, and the particular case of the integrable two-body problem. This is followed by the virial theorem and its applications. Space is dedicated to relaxation processes in N-body systems, before approaching the continuous limit and the corresponding field equations (Laplace's and Poisson's equations) together with their immediate consequences. Finally, Newton's classical theorems are demonstrated.

2.1 Discrete Mass Distributions

2.1.1 Single Particle Gravitational Potential

Let P be a point of mass m_P located at position $\mathbf{r}$ in a Cartesian reference frame whose origin is occupied by another point of mass m (Fig. 2.1).

The Newtonian gravity force exerted by the particle of mass m on the particle of mass m_P is

$$\mathbf{f}(\mathbf{r}) = -G\frac{m\,m_p}{r^2}\,\mathbf{e}_r = -G\frac{m\,m_p}{r^3}\,\mathbf{r}, \tag{2.1}$$

that, dividing by m_P, gives the force per unit mass (acceleration) acting on the particle

$$\mathbf{F}(\mathbf{r}) = -G\frac{m}{r^3}\,\mathbf{r}. \tag{2.2}$$

R. A. Capuzzo Dolcetta, *Classical Newtonian Gravity and Potential Theory*, UNITEXT for Physics, https://doi.org/10.1007/978-3-032-13036-5_2

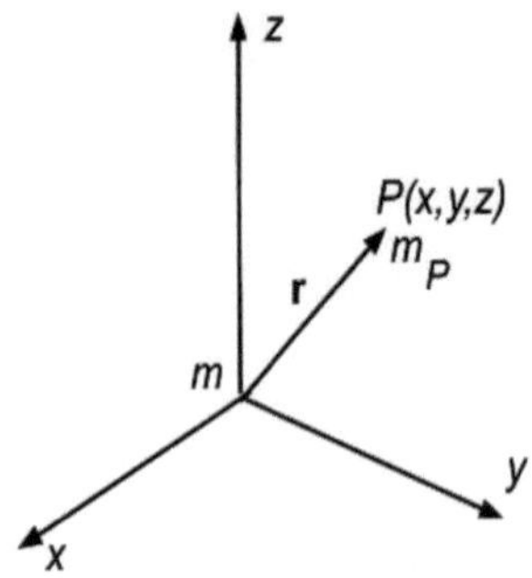

Fig. 2.1 A Cartesian reference frame centred on the particle of mass m. The particle in $P(x, y, z)$ feels the potential generated by m

Letting $U(r) = G(m/r)$, it follows that

$$\nabla U = \nabla G\frac{m}{r} = Gm\,\nabla\frac{1}{r} = -G\frac{m}{r^2}\nabla r = -G\frac{m}{r^2}\mathbf{e}_r = \mathbf{F}, \tag{2.3}$$

so that $U(r)$ is the *single particle* Newtonian potential.

Because, in $\mathbb{R}^n \setminus \{0\}$, $\nabla \cdot Gm\frac{\mathbf{r}}{r^3} = Gm\frac{3-n}{r^3}$, when $n = 3$ (i.e. in $\mathbb{R}^3$) we have

$$\nabla \cdot \mathbf{F} = \nabla \cdot \nabla U \equiv \nabla^2 U = 0. \tag{2.4}$$

This means that (i) $\mathbf{F}$ is a solenoidal vector field, and (ii) $U(\mathbf{r})$ is a harmonic function in the empty space around the particle m placed in the origin (a harmonic function U is a twice continuously differentiable solution of Laplace's equation $\nabla^2 U = 0$).

The vector equation of motion of the P particle with given initial conditions (i.c.) is

$$\begin{cases} \ddot{\mathbf{r}} = -G\dfrac{m}{r^3}\,\mathbf{r} \\ \mathbf{r}(0) = \mathbf{r}_0 \\ \dot{\mathbf{r}}(0) = \dot{\mathbf{r}}_0. \end{cases} \tag{2.5}$$

The set in Eq. 2.5 corresponds to a system of three scalar second-order ordinary differential equations. The motion of the P particle in the space around the particle of mass m located at the origin of the coordinate frame and providing the potential $G(m/r)$ is the typical case of *one-body* problem. Of course, due to Newton's third principle, the particle of mass m is also feeling the acceleration induced by the particle of mass m_P, and so the coordinate frame anchored in the m particle is *not* inertial. It becomes inertial in the limit $m \gg m_P$ (as, for instance, in the case of the Sun (m)-Earth (m_P) system).

Note that the 2nd order system in Eq. 2.5 can be easily reduced to a system of 6, 1st order, differential equations, by the introduction of the auxiliary vector $\mathbf{v} = \dot{\mathbf{r}}$. This is not relevant for the one-body (and two-body) point-like Newtonian cases because the explicit solution is known, while it is needed when the explicit solution does not exist and should be approximated by numerical methods.

2.1.2 *Motion of a Particle Subject to Gravity in a Resisting Medium*

The equation of motion of a particle of constant mass m_P submitted to both a conservative force $\mathbf{f}_c = m_P \nabla U$ and a dissipative force linearly dependent on the velocity, $\mathbf{f}_d = -\lambda \dot{\mathbf{r}}$ ($\lambda > 0$), is

$$\ddot{\mathbf{r}} = \frac{\mathbf{f}_c + \mathbf{f}_d}{m_P} = \nabla U - \frac{\lambda}{m_P}\dot{\mathbf{r}}. \tag{2.6}$$

In the case of a satellite orbiting the Earth, the dissipative force is caused by the atmospheric drag, and the coefficient λ is usually assumed as dependent on the satellite speed $\|\mathbf{v}\|$. For the purpose of an acceptable estimate of the drag force acting on the satellite, considered as having a non-zero cross-sectional area, it is usually assumed

$$\lambda = \frac{1}{2} C_D \rho A \|\mathbf{v}\|, \tag{2.7}$$

where ρ is the atmospheric mass density, decreasing exponentially with altitude (barometric law), $C_D > 0$ is a drag coefficient dependent on the satellite shape and surface characteristics, and A is the satellite cross-sectional area. In low Earth, quasi-circular, orbit, $\|\mathbf{v}\|$ is almost constant and equal to the local equilibrium centrifugal speed, slowly increasing along the satellite orbital decay. Anyway, for the sake of simplicity, in what follows we consider λ as independent of $\|\mathbf{v}\|$.

A scalar multiplication by $\dot{\mathbf{r}}$ of both sides of Eq. 2.6 followed by an integration over time leads to

$$\int_0^t \dot{\mathbf{r}} \cdot \ddot{\mathbf{r}}\, \mathrm{d}t = \int_0^t \dot{\mathbf{r}} \cdot \nabla U\, \mathrm{d}t - \frac{\lambda}{m_P} \int_0^t \dot{\mathbf{r}} \cdot \dot{\mathbf{r}}\, \mathrm{d}t, \tag{2.8}$$

from which

$$\frac{1}{2}\dot{\mathbf{r}} \cdot \dot{\mathbf{r}} - U(\mathbf{r}) = \frac{1}{2}\dot{\mathbf{r}}_0 \cdot \dot{\mathbf{r}}_0 - U(\mathbf{r}_0) - \frac{\lambda}{m_P} \int_0^t \dot{\mathbf{r}} \cdot \dot{\mathbf{r}}\, \mathrm{d}t, \tag{2.9}$$

where $\mathbf{r}_0 = \mathbf{r}(0)$ and $\dot{\mathbf{r}}_0 = \dot{\mathbf{r}}(0)$.

Defining the mechanical energy per unit mass, E, as $E = \frac{1}{2}\dot{\mathbf{r}} \cdot \dot{\mathbf{r}} - U(\mathbf{r})$, we can write the above relation as

$$E(t) = E_0 - \frac{\lambda}{m_P} \int_0^t \dot{\mathbf{r}} \cdot \dot{\mathbf{r}} \, \mathrm{d}t. \tag{2.10}$$

Let us note that

$$L_d = -\frac{\lambda}{m_P} \int_0^t \dot{\mathbf{r}} \cdot \dot{\mathbf{r}} \, \mathrm{d}t = -\int_{\mathbf{r}(0)}^{\mathbf{r}(t)} \frac{\lambda}{m_P} \dot{\mathbf{r}} \cdot \mathrm{d}\mathbf{r} = \int_{\mathbf{r}(0)}^{\mathbf{r}(t)} \frac{\mathbf{f}_d}{m_P} \cdot \mathrm{d}\mathbf{r} \leq 0, \tag{2.11}$$

represents the (always non positive) work (per unit mass) done by the dissipative force $\mathbf{f}_d$ in the time interval $[0, t]$. Consequently the expression

$$\dot{E} = -\frac{\lambda}{m_P} \dot{\mathbf{r}} \cdot \dot{\mathbf{r}} = \frac{\mathbf{f}_d}{m_P} \cdot \mathbf{v} \leq 0 \tag{2.12}$$

gives the power being dissipated at any time t.

Note 10

Let us recall a concept of differential calculus.

Definition 2.1 Let A be an open set in $\mathbb{R}^3$ and $Xdx+Ydy+Zdz$ a differential form, where X, Y, Z are continuous and differentiable functions of x, y, z in A. The form is called *exact* if there exists a function $g(x, y, z)$ such that

$$\mathrm{d}g = X\mathrm{d}x + Y\mathrm{d}y + Z\mathrm{d}z,$$

where $\mathrm{d}g$ is the total differential of g.

Given that, we have this theorem:

Theorem 2.1 *The differential form $Xdx + Ydy + Zdz$, where X, Y, Z are continuous and differentiable functions of x, y, z in a simply connected set A, is an exact differential if and only if the following equalities hold*

$$\frac{\partial X}{\partial y} = \frac{\partial Y}{\partial x}, \qquad \frac{\partial X}{\partial z} = \frac{\partial Z}{\partial x}, \qquad \frac{\partial Y}{\partial z} = \frac{\partial Z}{\partial y}. \tag{2.13}$$

Going back to what said above on the motion of a particle in presence of a dissipative force, to recover a *constant* of motion[a] in the case of dissipative force of the type in Eq. 2.6, it is necessary that $\nabla \wedge \dot{\mathbf{r}} = 0$, which corresponds

(continued)

Theorem 2.1 (continued)
to a series of equalities similar to those in 2.13:

$$\frac{\partial \dot{x}}{\partial y} = \frac{\partial \dot{y}}{\partial x}, \qquad \frac{\partial \dot{x}}{\partial z} = \frac{\partial \dot{z}}{\partial x}, \qquad \frac{\partial \dot{y}}{\partial z} = \frac{\partial \dot{z}}{\partial y}. \tag{2.14}$$

These relations would imply (by the previous theorem) the existence of a scalar function $\Phi(\mathbf{r})$ whose total differential is the differential form $\dot{x}dx + \dot{y}dy + \dot{z}dz$ and thus

$$\dot{\mathbf{r}} = \nabla\Phi(\mathbf{r}), \tag{2.15}$$

which would imply

$$-\frac{\lambda}{m_P}\int_{\mathbf{r}(0)}^{\mathbf{r}(t)} \nabla\Phi \cdot \mathrm{d}\mathbf{r} = -\frac{\lambda}{m_P}\left[\Phi(\mathbf{r}) - \Phi(\mathbf{r}_0)\right], \tag{2.16}$$

yielding a new, allegedly generalized, "energy" constant of motion

$$E + \frac{\lambda}{m_P}\Phi(\mathbf{r}) = E_0 + \frac{\lambda}{m_P}\Phi(\mathbf{r}_0) = \text{const.} \tag{2.17}$$

Anyway, the questions are:

- does this make sense?
- what is its physical meaning?

The answers are left to the reader.

---

[a] A constant of motion is a function of *phase-space* coordinates and time that remains constant along every trajectory of motion; an integral of motion is a particular constant of motion that depends only upon phase-space coordinates and not upon time. The phase space is a 6-dimensional space where every point (x, y, z, v_x, v_y, v_z) represents position and velocity of a particle.

2.1.2.1 The Case of Variable Mass Satellite

Assuming $\nabla U = G(m_E/r)$ in Eq. 2.6 makes it the simplest modelization of a point-like satellite of constant mass m_P moving around the Earth approximated as a spherically symmetric distribution of matter of total mass m_E. To be integrated, Eq. 2.6 must be accompanied by i.c. $\mathbf{r}(0)$ and $\dot{\mathbf{r}}(0)$.

In the case of a point-like satellite of variable mass $m_P(t)$, after defining the linear momentum $\mathbf{q} = m_P\dot{\mathbf{r}}$, the Newtonian equation of motion 2.6 is replaced by

$$\dot{\mathbf{q}} = m_P\ddot{\mathbf{r}} + \dot{m}_P\dot{\mathbf{r}} = m_P\nabla U - \lambda\dot{\mathbf{r}}, \tag{2.18}$$

equivalent to

$$\ddot{\mathbf{r}} = \nabla U - \frac{\lambda}{m_P}\dot{\mathbf{r}} - \frac{\dot{m}_P}{m_P}\dot{\mathbf{r}} = \nabla U - \frac{\lambda + \dot{m}_P}{m_P}\dot{\mathbf{r}}. \tag{2.19}$$

The standard case is that of $\dot{m}_P < 0$, due to fuel consumption of the on-board rocket engine. This implies, even in the case of isotropic mass loss, a reduction of the drag force acting on the satellite that makes the shrinking of its orbit.

2.1.3 The Gravitational N-Body Case

Given a set of isolated N particles subjected to their mutual pairwise interaction force, the force acting on the ith particle, $\mathbf{f}_i$ $(i = 1, 2, \ldots, N)$, is given by the sum over $j = 1, 2, \ldots, N$ (excluding $j = i$) of the force contributed by every other jth particle, $\mathbf{f}_{ij}$. In the case of purely Newtonian interaction, if we indicate with $\mathbf{r}_i$ the position vector of the ith particle and with m_i its mass, the force exerted by the jth particle of mass m_j on the ith is, $\forall i \neq j$,

$$\mathbf{f}_{ij} = f_{ij}\frac{\mathbf{r}_j - \mathbf{r}_i}{\|\mathbf{r}_j - \mathbf{r}_i\|} \quad \text{with} \quad f_{ij} = G\frac{m_j m_i}{\|\mathbf{r}_j - \mathbf{r}_i\|^2}, \tag{2.20}$$

so that

$$\mathbf{f}_i = \sum_{\substack{j=1 \\ j\neq i}}^{N} \mathbf{f}_{ij} = \sum_{\substack{j=1 \\ j\neq i}}^{N} G\frac{m_j m_i}{\|\mathbf{r}_j - \mathbf{r}_i\|^3}(\mathbf{r}_j - \mathbf{r}_i), \tag{2.21}$$

is the total force exerted by the other $N - 1$ bodies of the system on the ith. Clearly,

$$\mathbf{f}_{ij} = -\mathbf{f}_{ji}, \qquad f_{ij} = f_{ji}. \tag{2.22}$$

If we define the total linear momentum of the system as $\mathbf{q} = \sum\limits_{i=1}^{N} m_i\dot{\mathbf{r}}_i$, in absence of external forces the relations (2.22) imply its conservation. This because

$$\dot{\mathbf{q}} = \sum_{i=1}^{N} m_i\ddot{\mathbf{r}}_i = \mathbf{f}^{(int)} + \mathbf{f}^{(ext)}, \tag{2.23}$$

where $\mathbf{f}^{(int)} = \sum_{i=1}^{N} \mathbf{f}_i$ is the resultant of the *internal* forces on the system, which is zero due to Eq. 2.22, while $\mathbf{f}^{(ext)}$ is the resultant of the *external* forces on the system, surely zero if the system is isolated. If $\mathbf{f}^{(ext)} = 0$, Eq. 2.23 implies linear momentum conservation

$$\mathbf{q} = \sum_{i=1}^{N} m_i \dot{\mathbf{r}}_i = \mathbf{q}_0 = const.. \tag{2.24}$$

The relation above corresponds to a vector, *classical*, integral of motion, i.e. $\mathbf{q}$ is a vector which is constant along all possible solutions of the equation of motion and *does not* depend on t.

If we integrate over time the lhs and rhs of Eq. 2.24 we get

$$\int_{t_0}^{t} \mathbf{q}\, \mathrm{d}t = \mathbf{q}_0(t - t_0), \tag{2.25}$$

which leads to the vector, trivial, *constant* of motion

$$\sum_{i=1}^{N} m_i \mathbf{r}_i(t) = \sum_{i=1}^{N} m_i \mathbf{r}_i(t_0) + \mathbf{q}_0(t - t_0). \tag{2.26}$$

Accounting for the definition of centre of mass position $\mathbf{r}_G = \sum_{i=1}^{N} \mathbf{r}_i / M$, where $M = \sum_{i=1}^{N} m_i$ is the total mass of the system, the expression above can be also written as

$$\mathbf{r}_G(t) = \mathbf{r}_G(t_0) + \frac{\mathbf{q}_0}{M}(t - t_0), \tag{2.27}$$

which represents the linear translation in time of the centre of mass.

On the other hand, the system respects also the 2nd fundamental equation of the dynamics, which states that $\dot{\mathbf{l}} = \boldsymbol{\tau}$, where

$$\mathbf{l} = \sum_{i=1}^{N} \mathbf{r}_i \wedge m_i \dot{\mathbf{r}}_i, \tag{2.28}$$

is the total angular momentum of the system and

$$\boldsymbol{\tau} = \boldsymbol{\tau}^{(int)} + \boldsymbol{\tau}^{(ext)} = \sum_{i=1}^{N} \mathbf{r}_i \wedge \mathbf{f}_i^{(int)} + \sum_{i=1}^{N} \mathbf{r}_i \wedge \mathbf{f}_i^{(ext)}, \tag{2.29}$$

is the total torque acting on the system, due to internal (pairwise, $\mathbf{f}_i^{(int)} = \mathbf{f}_i$ where $\mathbf{f}_i$ comes from Eq. 2.21) and possible external forces.

Now, if we consider the system as composed of the whole set of pairs (i,j), the moment of internal forces, $\boldsymbol{\tau}^{(int)}$, can be written as the sum of the contribution of all these (distinct) pairs, which are, in number, $\binom{N}{2} = \frac{N(N-1)}{2}$, so that

$$\boldsymbol{\tau}^{(int)} = \sum_{i=1}^{N} \mathbf{r}_i \wedge \sum_{\substack{j=1 \\ j\neq i}}^{N} \mathbf{f}_{ij} = \sum_{i=1}^{N} \sum_{\substack{j=1 \\ j\neq i}}^{N} \mathbf{r}_i \wedge \mathbf{f}_{ij} = \sum_{\text{pairs}} (\mathbf{r}_i \wedge \mathbf{f}_{ij} + \mathbf{r}_j \wedge \mathbf{f}_{ji}) =$$

$$= \sum_{\text{pairs}} (\mathbf{r}_i - \mathbf{r}_j) \wedge G \frac{m_i m_j}{\|\mathbf{r}_i - \mathbf{r}_j\|^3} (\mathbf{r}_j - \mathbf{r}_i) = 0, \qquad \forall i, j. \tag{2.30}$$

Consequently, the second fundamental equation of dynamics reduces to

$$\dot{\mathbf{l}} = \boldsymbol{\tau}^{(ext)}. \tag{2.31}$$

This means that, for an N-body system which is either isolated or submitted to a spherically symmetric external force field, for which $\boldsymbol{\tau}^{(ext)} = 0$, the total angular momentum is another vector integral of motion, $\mathbf{l} = \mathbf{l}_0 = const.$.

So far, we have determined two vector integral of motions ($\mathbf{q}$ and $\mathbf{l}$) and one vectorl constant of motion ($\mathbf{r}_G(t)$). This means a total of 9 scalar functions which link the phase-space coordinates of the N bodies, independently of N.

In the following section we determine the last classical integral of motion available, that of the mechanical energy.

2.1.3.1 Potential and Energy of N Gravitating Bodies

Given N points whose masses and positions are m_k and $\mathbf{r}_k$, $k = 1, 2, \ldots, N$, if they interact via Newtonian gravity only, the function defined as

$$\widetilde{U}(\mathbf{r}_1, \cdots, \mathbf{r}_N) = \frac{G}{2} \sum_{\substack{(i,j)=1 \\ i\neq j}}^{N} \frac{m_i m_j}{\|\mathbf{r}_i - \mathbf{r}_j\|} = G \sum_{i=1}^{N-1} \sum_{j=i+1}^{N} \frac{m_i m_j}{r_{ij}} =$$

$$= G \sum_{j=1}^{N-1} \sum_{i=j+1}^{N} \frac{m_i m_j}{r_{ij}}, \tag{2.32}$$

(where we have made use of the fact that $r_{ij} \equiv \|\mathbf{r}_i - \mathbf{r}_j\| = r_{ji}$) is, with sign convention adopted in this book, the opposite of the *gravitational potential energy*, $\Omega = -\widetilde{U} < 0$.[1]

Given Eq. 2.32, it results

$$\mathbf{f}_k = \nabla_k \widetilde{U}, \qquad \text{with} \qquad \nabla_k \equiv \left(\frac{\partial}{\partial x_k}, \frac{\partial}{\partial y_k}, \frac{\partial}{\partial z_k} \right). \tag{2.33}$$

This is easy to verify; indeed

$$\frac{\partial \widetilde{U}}{\partial x_k} = -\frac{G}{2} \sum_{\substack{(i,j)=1 \\ i \neq j}}^{N} \frac{m_i m_j}{r_{ij}^2} \frac{\partial r_{ij}}{\partial x_k} = -\frac{G}{2} \sum_{\substack{i=1 \\ i \neq k}}^{N} \frac{m_i m_k}{r_{ik}^2} \frac{\partial r_{ik}}{\partial x_k} - \frac{G}{2} \sum_{\substack{j=1 \\ j \neq k}}^{N} \frac{m_k m_j}{r_{kj}^2} \frac{\partial r_{kj}}{\partial x_k} =$$

$$= -G \sum_{\substack{l=1 \\ l \neq k}}^{N} \frac{m_k m_l}{r_{kl}^3} (x_k - x_l) = f_{kx}, \tag{2.34}$$

and similarly for the y_k and z_k coordinates.[2] Given this potential, the equations of motion of the generic ith body of the N-body system are synthetically written as

$$\begin{cases} \ddot{\mathbf{r}}_i = \dfrac{\nabla_i \widetilde{U}}{m_i} \\ \mathbf{r}_i(0) = \mathbf{r}_{i0} \\ \dot{\mathbf{r}}_i(0) = \dot{\mathbf{r}}_{i0}, \end{cases} \tag{2.35}$$

where the initial conditions for both position and velocity have been specified.

The particular case of a single ith *test* particle of mass m_i, identified by its position vector $\mathbf{r}_i$ and moving (without influencing the motion of the other bodies) in the force field generated by the other $N-1$ bodies, leads to the following equation of motion

$$\ddot{\mathbf{r}}_i = \sum_{j=1}^{N} G \frac{m_j}{r_{ij}^3} (\mathbf{r}_j - \mathbf{r}_i), \tag{2.36}$$

[1] Note that $\widetilde{U}$ in Eq. 2.32 is NOT per unit mass.

[2] Note the antisymmetry of the action of ∇_k with respect to the generic (k, j)-pair distance; actually

$$\nabla_k \|\mathbf{r}_k - \mathbf{r}_j\| = \frac{\mathbf{r}_k - \mathbf{r}_j}{\|\mathbf{r}_k - \mathbf{r}_j\|} = -\nabla_j \|\mathbf{r}_k - \mathbf{r}_j\|.$$

This means that $\nabla_k r_{kj}$ is the unit vector pointing from the jth to the kth object; conversely $\nabla_j r_{kj}$ is the unit vector pointing from the kth to the jth object.

which corresponds to this potential

$$\widetilde{U} = \sum_{\substack{j=1 \\ j \neq i}}^{N} G \frac{m_i m_j}{r_{ij}}. \tag{2.37}$$

2.1.3.2 Mechanical Energy of the N Bodies

Given the ith body's equation of motion in an N-body system

$$\ddot{\mathbf{r}}_i = \frac{\nabla_i \widetilde{U}}{m_i} = \sum_{\substack{j=1 \\ j \neq i}}^{N} G \frac{m_j}{\|\mathbf{r}_j - \mathbf{r}_i\|^3} (\mathbf{r}_j - \mathbf{r}_i), \tag{2.38}$$

where $\widetilde{U}$ is the function given by Eq. 2.32, a summation over $i = 1, 2, \ldots, N$ followed by a scalar multiplication by $\dot{\mathbf{r}}_i$ of the left and right sides, an integration over t from t_0 to t and a further summation over $i = 1, 2, \ldots, N$, leads eventually to

$$\begin{aligned}
&\int_{t_0}^{t} \sum_{i=1}^{N} m_i \ddot{\mathbf{r}}_i \cdot \dot{\mathbf{r}}_i \mathrm{d}t = \int_{t_0}^{t} \sum_{i=1}^{N} m_i \dot{\mathbf{r}}_i \cdot \mathrm{d}\dot{\mathbf{r}}_i = \sum_{i=1}^{N} \left(\frac{1}{2} m_i \dot{\mathbf{r}}_i \cdot \dot{\mathbf{r}}_i - \frac{1}{2} m_i \dot{\mathbf{r}}_{i0} \cdot \dot{\mathbf{r}}_{i0} \right) = \\
&= \int_{t_0}^{t} \sum_{i=1}^{N} \nabla_i \widetilde{U} \cdot \dot{\mathbf{r}}_i \mathrm{d}t = \int_{t_0}^{t} \sum_{i=1}^{N} \left(\frac{\partial \widetilde{U}}{\partial x_i} \mathrm{d}x_i + \frac{\partial \widetilde{U}}{\partial y_i} \mathrm{d}y_i + \frac{\partial \widetilde{U}}{\partial z_i} \mathrm{d}z_i \right) = \int_{t_0}^{t} \mathrm{d}\widetilde{U} = \\
&= \widetilde{U}(\mathbf{r}_1(t), \mathbf{r}_2(t), \ldots, \mathbf{r}_N(t)) - \widetilde{U}(\mathbf{r}_{10}, \mathbf{r}_{20}, \ldots, \mathbf{r}_{N0}).
\end{aligned} \tag{2.39}$$

A consequence of the above relations is the conservation of the *total* mechanical energy of the N-body system

$$\begin{aligned}
\widetilde{E} &= \sum_{i=1}^{N} \frac{1}{2} m_i \dot{\mathbf{r}}_i \cdot \dot{\mathbf{r}}_i - \widetilde{U}(\mathbf{r}_1(t), \mathbf{r}_2(t), \ldots, \mathbf{r}_N(t)) = \widetilde{T} - \widetilde{U} = \widetilde{T} + \Omega = \\
&= \sum_{i=1}^{N} \frac{1}{2} m_i \dot{\mathbf{r}}_{i0} \cdot \dot{\mathbf{r}}_{i0} - \widetilde{U}(\mathbf{r}_{10}, \mathbf{r}_{20}, \ldots, \mathbf{r}_{N0}) = \widetilde{E}_0 = const.
\end{aligned} \tag{2.40}$$

where $\widetilde{T}$ is the total kinetic energy.

Equation 2.40 shows that $\widetilde{E}$ is a scalar integral of motion. There are no other classical integrals and constants of motion in a classical Newtonian N-body system, so the total number of scalar integrals and constants of motion is 10, regardless of N.

Table 2.1 Summary of conservation properties of the N-body problem

Symmetry	Conserved quantity	Number of c.q.
Time translation	Mechanical energy	1
Space translation	Linear momentum	3
Space rotation	Angular momentum	3
Galilean	Centre of mass velocity	3

The existence of 10 constants of motion was a priori-guaranteed by the validity of the Noether's theorem, published in 1918, when applied to the gravitational, Newtonian, N-body case. Noether's theorem essentially says that to every continuous symmetry of a physical law, representable as an invariance of the law under a specific transformation, it corresponds the existence of a conserved physical quantity. In the case of the classical equations of motion of a system of particles interacting via Newtonian potential, there is invariance in time, in space translation and rotation. Additionally, there is the so-called *Galilean* invariance, corresponding to invariance respect the change of reference in the class of inertial frames. Table 2.1 summarizes these invariances and the corresponding conserved quantities.

This limited number of constraints in phase-space explains why only for $N = 2$ the problem is fully *integrable*. In 2-body case, indeed, the trajectory is found to be a conic section (ellipse, parabola or hyperbola), while the position along the trajectory requires the numerical solution of a non-linear equation, called *Kepler's equation*. Already for $N = 3$ explicit solutions are available only in particular cases.

It is relevant noting that only the *total* energy $\widetilde{E}$ is conserved, while the individual particle energy $\widetilde{E}_i = \frac{1}{2}m_i v_i^2 - G\sum_{\substack{j=1\\ j\neq i}}^{N} \frac{m_i m_j}{r_{ij}}$ is *not*. This means that, even if during the system evolution a particle assumes a positive energy, $\widetilde{E}_i > 0$, which corresponds to a speed larger than the local escape velocity, it is not guaranteed it will be "expelled" from the N-body system because it could interact with other field objects and reduce $\widetilde{E}_i$ to a negative value again. This is of course unlikely whenever the speed over the local escape velocity is maintained in a low-density environment.

Historical Note

In 1884 the Swedish mathematician Gösta Mittag–Leffler (1846–1927) proposed to the king Oscar II of Sweden and Norway (1829–1907) to award a prize of a gold medal and 2500 kr for a solution to an unsolved mathematical problem among four proposed by a committee. Note that 2500 kr was an amount a bit larger than Mittlag-Leffler annual salary as professor. The committee was composed by Mittag-Leffler, Charles Hermite (1822–1901) and Karl Weierstrass (1815–1897) and the answers were due within June 1st 1888. The four questions were officially announced in volume 7 of Acta Mathematica (journal established by Mittag-Leffler in 1882) in the middle

(continued)

of 1885. The first question (which attracted the attention of 5 out of 12 applicants) was actually concerned with the classical N-body problem:

For a system of arbitrarily many mass points that attract each other according to Newton's law, assuming that no two points ever collide, find a series expansion of the coordinates of each point in known functions of time converging uniformly for any period of time.

Henry Poincaré (1854–1912) was awarded the prize for his answer to the above question, although his answer was not a rigorous solution of the problem posed. The committee, actually, appreciated the novelty of his approach to the topic, based on introduction of a new geometrical and topological vision of the problem that impressed Weierstrass. As a matter of fact, the prize was awarded to Poincaré because his heavy work on the N-body problem opened a new era in the history of celestial mechanics. His work was fundamental also for the development of the general theory of dynamical systems. It is unanimously acknowledged that Poincaré's work corresponded to the birth of the concept of *deterministic chaos.*

A remarkable epilogue to king Oscar's prize competition occurred when—about twenty years after the prize was awarded—the Finnish mathematician and astronomer Karl F. Sundman (1873–1949) published articles where he gave a proof of the convergence of an infinite series solution to the three-body problem for almost all initial conditions, basing on classical arguments and results. In spite of its theoretical interest, the Sundman's result is of no practical use because of the exceedingly slow convergence of the series solution. It is worth noting that the Sundman's result was extended to the general N-body case by Qiu-dong Wang in 1991 [12].

2.1.3.3 The Scalar Virial Theorem

An important relation involving the location in the *phase space* of a system of N interacting objects takes the name of *virial* theorem.

Given a set of N point masses which in a Cartesian reference frame are identified by position vectors $\mathbf{r}_i$, $i = 1, 2, \ldots, N$, their moment of inertia with respect to the origin, $\widetilde{I}$, is

$$\widetilde{I} = \sum_{i=1}^{N} m_i \mathbf{r}_i \cdot \mathbf{r}_i. \tag{2.41}$$

Taking its second derivative about the time we have

$$\ddot{\widetilde{I}} = 2 \sum_{i=1}^{N} m_i (\dot{\mathbf{r}}_i \cdot \dot{\mathbf{r}}_i + \mathbf{r}_i \cdot \ddot{\mathbf{r}}_i), \tag{2.42}$$

which may be written as

$$\frac{1}{2}\ddot{\widetilde{I}} = 2\widetilde{T} + \sum_{i=1}^{N} \mathbf{r}_i \cdot \nabla_i \widetilde{U}. \tag{2.43}$$

The quantity

$$\sum_{i=1}^{N} \mathbf{r}_i \cdot \nabla_i \widetilde{U} = \sum_{i=1}^{N} \left(x_i \frac{\partial \widetilde{U}}{\partial x_i} + y_i \frac{\partial \widetilde{U}}{\partial y_i} + z_i \frac{\partial \widetilde{U}}{\partial z_i} \right), \tag{2.44}$$

is called *Clausius' virial*,[3] after Rudolf Clausius (1822-1888), who introduced the so-called virial theorem in 1870.

Let us recall an important mathematical aspect.

Definition 2.2 Let $f(\mathbf{r})$ be a real function of m real variables; this function is called *homogeneous* of degree n if for any $\alpha \neq 0$ the relation

$$f(\alpha \mathbf{r}) = \alpha^n f(\mathbf{r}) \qquad \forall \mathbf{r} \in \mathbb{R}^m, \tag{2.45}$$

holds. When Eq. 2.45 is valid for $\alpha > 0$ or $\alpha < 0$ we speak of $f(\mathbf{r})$ as *positively* or *negatively* homogeneous, respectively.

Lemma 2.1 *If the interaction potential $\widetilde{U}(\mathbf{r}_1, \cdots, \mathbf{r}_N)$ is a homogeneous function of degree n, it results*

$$\sum_{i=1}^{N} \mathbf{r}_i \cdot \nabla_i \widetilde{U} = n\widetilde{U}. \tag{2.46}$$

Proof If the function $\widetilde{U}(\mathbf{r}_1, \cdots, \mathbf{r}_N)$ is homogeneous of degree n it satisfies the relation 2.45, and so, by differentiation with respect to α of both sides of that equation, we have

$$\frac{\partial}{\partial \alpha}\widetilde{U}(\alpha\mathbf{r}_1, \cdots, \alpha\mathbf{r}_N) = \frac{\partial}{\partial \alpha}\left[\alpha^n \widetilde{U}(\mathbf{r}_1, \cdots, \mathbf{r}_N)\right], \tag{2.47}$$

that develops into

$$\sum_{i=1}^{N} \left(\frac{\partial \widetilde{U}}{\partial \alpha x_i} \frac{\partial \alpha x_i}{\partial \alpha} + \frac{\partial \widetilde{U}}{\partial \alpha y_i} \frac{\partial \alpha y_i}{\partial \alpha} + \frac{\partial \widetilde{U}}{\partial \alpha z_i} \frac{\partial \alpha z_i}{\partial \alpha} \right) = n\alpha^{n-1}\widetilde{U}, \tag{2.48}$$

equivalent to

[3] The term "virial" comes from the Latin *vis* (force).

$$\sum_{i=1}^{N}(\nabla_{\alpha\mathbf{r}_i}\widetilde{U})\cdot\mathbf{r}_i = n\alpha^{n-1}\widetilde{U}. \tag{2.49}$$

Given the arbitrariness of α we can set $\alpha = 1$ in the above equation and so the relation 2.46 is obtained.

Thanks to Lemma 2.1 we can write Eq. 2.43 as

$$\frac{1}{2}\ddot{\widetilde{I}} = 2\widetilde{T} + n\widetilde{U}, \tag{2.50}$$

which is the actual thesis of the virial *scalar* theorem.

Due to that the opposite of the potential energy of a set of N particles interacting via Newtonian classical gravity is given by (Eq. 2.32)

$$\widetilde{U} = \frac{G}{2}\sum_{\substack{(i,j)=1\\ i\neq j}}^{N}\frac{m_i m_j}{r_{ij}}, \tag{2.51}$$

which is (as easily seen) a positively homogeneous function of degree $n = -1$, the virial theorem thesis (2.50) assumes the form

$$\frac{1}{2}\ddot{\widetilde{I}} = 2\widetilde{T} - \widetilde{U} = 2\widetilde{T} + \Omega = \widetilde{T} + \widetilde{E}, \tag{2.52}$$

where $\Omega \equiv -\widetilde{U}$ is the potential energy of the system. Note that if $\ddot{\widetilde{I}} = 0$ then $\widetilde{E} = -\widetilde{T}$, that means $d\widetilde{E}/d\widetilde{T} = -1$: the heat capacity of a self-gravitating system is negative!

We limit here to noting that the virial scalar theorem is a particular case of the more general virial *tensor* theorem, which refers to the 3×3 inertia tensor

$$\widetilde{I}_{ik} = \sum_{j=1}^{N} m_j x_{j,i} x_{j,k}, \tag{2.53}$$

where the convention to indicate Cartesian coordinate of the jth particle $x_{j,1} = x_j, x_{j,2} = y_j, x_{j,3} = z_j$ is adopted. Clearly, the moment of inertia $\widetilde{I}$ defined in Eq. 2.41 is the trace of the inertia tensor.

2.1.3.4 Consequences of the Virial Theorem

By dividing the 2 sides of the virial relation (Eq. 2.50, where we set $n = -1$) by $|\Omega| = -\Omega$ we have

$$\frac{1}{2}\frac{\ddot{\tilde{I}}}{|\Omega|} = \frac{2\tilde{T}}{|\Omega|} - 1, \tag{2.54}$$

which, upon definition of the dimensionless *virial ratio* as $Q \equiv 2\tilde{T}/|\Omega| \geq 0$, translates into

$$Q = \frac{1}{2}\frac{\ddot{\tilde{I}}}{|\Omega|} + 1. \tag{2.55}$$

A system is said *virialized* if it is globally steady, i.e. if $\ddot{\tilde{I}} = 0$, and so $Q = 1\ \forall t$. It is *subvirial* when $Q < 1$, which corresponds to a negative acceleration of the inertia moment, $\ddot{\tilde{I}} < 0$ (initial contraction), and *supervirial* when $Q > 1$, which corresponds to a positive acceleration of the inertia moment, $\ddot{\tilde{I}} > 0$ (initial expansion).

If the system is virialized, $Q = 2\tilde{T}/|\Omega| = 1$, it has a root mean square velocity $v_{rms} = \sqrt{|\Omega|/M}$, where $M = \sum\limits_{i=1}^{N} m_i$ is the total system mass. The precise value of v_{rms} depends, of course, on the instantaneous value of Ω and so on the actual spatial distribution of the N bodies, but a good approximation is given by $|\Omega| = 3GM^2/(5R)$, in the assumption of a homogeneous distribution within a sphere of radius R (see Exercise 2.5 in Sect. 2.2.3). We call *virial* velocity, v_{vir}, the v_{rms} evaluated with the above approximation for Ω, and call *crossing* time the time needed to an object moving with the virial speed to entirely cross the system, $t_{cr} = 2R/v_{vir}$.

Let us now examine some particular cases concerning $\ddot{\tilde{I}} \neq 0$.

If the system motion is periodic with period P, performing an integral time average of Eq. 2.52 over P we get

$$\frac{1}{2}\langle\ddot{\tilde{I}}\rangle_P = \frac{1}{2}\frac{1}{P}\int\limits_0^P \ddot{\tilde{I}}\,\mathrm{d}t = \frac{1}{2}\frac{1}{P}\left[\dot{\tilde{I}}(P) - \dot{\tilde{I}}(0)\right] = 0 = 2\langle\tilde{T}\rangle_P + \langle\Omega\rangle_P. \tag{2.56}$$

In the above equation, we made use of that, if $\tilde{I}(t)$ is periodic, also $\dot{\tilde{I}}(t)$ is periodic. This is a general property. Indeed, to say that $\tilde{I}(t)$ is periodic means that $\tilde{I}(t+P) = \tilde{I}(t)$ for every t. Consequently

$$\frac{1}{2}\langle\ddot{\tilde{I}}\rangle_\tau = \frac{1}{2}\frac{1}{\tau}\int\limits_0^\tau \ddot{\tilde{I}}\,\mathrm{d}t = \frac{1}{2}\frac{1}{\tau}\left[\dot{I}(\tau) - \dot{\tilde{I}}(0)\right] = 2\langle\tilde{T}\rangle_\tau + \langle\Omega\rangle_\tau. \tag{2.57}$$

Thus, if τ is properly chosen of the order of the oscillation time of the system, it is $\dot{\tilde{I}}(\tau) \simeq \dot{\tilde{I}}(0)$ or, equivalently, $< Q >_\tau \simeq 1$, i.e. the system is globally virialized over the time scale τ.

In the case of a motion which is non-periodic but limited in the phase space, i.e.

$$\exists L, M > 0 : \quad r_i(t) \leq L, \quad \|\dot{\mathbf{r}}_i\| \leq M, \qquad \forall i \text{ and } \forall t, \tag{2.58}$$

it follows that $\dot{\widetilde{I}}$ is limited, i.e. $\exists H > 0$ such that $|\dot{\widetilde{I}}| \leq H, \quad \forall t$ and so

$$\begin{aligned} \frac{1}{2}\langle\ddot{\widetilde{I}}\rangle_\infty &= \frac{1}{2}\lim_{\tau\to\infty}\frac{1}{\tau}\left[\dot{\widetilde{I}}(\tau) - \dot{\widetilde{I}}(0)\right] \leq \\ &\leq \frac{1}{2}\lim_{\tau\to\infty}\frac{1}{\tau}\left[|\dot{\widetilde{I}}(\tau)| + |\dot{\widetilde{I}}(0)|\right] \leq \frac{1}{2}\lim_{\tau\to\infty}\frac{2H}{\tau} = 0. \end{aligned} \tag{2.59}$$

A last, but relevant, result is that if the system energy is positive, $\widetilde{E} > 0$, the system is *unstable* (unbound in physical space).

Actually

$$\frac{1}{2}\ddot{\widetilde{I}} = \widetilde{E} + \widetilde{T} \geq \widetilde{E} > 0, \tag{2.60}$$

and, by two integrations over time from 0 to t

$$\widetilde{I}(t) \geq \widetilde{E}t^2 + \dot{\widetilde{I}}(0)t + \widetilde{I}(0). \tag{2.61}$$

Given that $\widetilde{E} > 0$, the latter formula implies

$$\lim_{t\to\infty} \widetilde{I}(t) = +\infty, \tag{2.62}$$

that means the system is indeed *unstable*, in what exists at least an escaping particle, for which $r_i(t) \to \infty$.

2.1.3.5 The Role of an External Force

If the generic ith star of the set of N mutually gravitating bodies is also subjected to an *external* force field $\mathbf{f}_i^{(ext)}$, the virial expression (see Eq. 2.42, and Eq. 2.50 with $n = -1$) modifies as

$$\frac{1}{2}\ddot{\widetilde{I}} = 2\widetilde{T} - \widetilde{U} + \sum_{i=1}^{N} \mathbf{r}_i \cdot \mathbf{f}_i^{(ext)}, \tag{2.63}$$

and so virialization is achieved when

$$\frac{1}{2}\ddot{\widetilde{I}} = 2\widetilde{T} - \widetilde{U} + \sum_{i=1}^{N} \mathbf{r}_i \cdot \mathbf{f}_i^{(ext)} = 0, \tag{2.64}$$

that is $2\widetilde{T} + \Omega_{tot} = 0$, where

$$\Omega_{tot} = -\widetilde{U} + \sum_{i=1}^{N} \mathbf{r}_i \cdot \mathbf{f}_i^{(ext)}, \tag{2.65}$$

is an effective gravitational potential energy term.

In the particular case of a smooth external force field radially pointing to the origin,[4] expressed as

$$\mathbf{f}_i^{(ext)} = -f_i^{(ext)} \frac{\mathbf{r}_i}{r_i}, \tag{2.66}$$

the external field contribution to the inertia moment in Eq. 2.63 is negative. Defining $Q_{tot} = \ddot{\widetilde{I}}/(2|\Omega_{tot}|) + 1$, virialization corresponds again to $Q_{tot} = 1$. On the other side, the virial ratio $Q = 2\widetilde{T}/|\Omega|$, which is now written as

$$Q = \frac{1}{2}\frac{\ddot{\widetilde{I}}}{|\Omega|} + 1 + \frac{\sum_{i=1}^{N} f_i^{(ext)} r_i}{|\Omega|} \tag{2.67}$$

is, at virialization, $Q > 1$.[5]

2.1.3.6 Crucial Aspects of the N-Body Problem

The system of equations of motion 2.35, also reported and discussed in Appendix A, in spite of its apparent simplicity is extremely difficult to solve, for many reasons. These reasons are connected to the apparently simple Newtonian pair-interaction potential $U_{ij} \propto 1/r_{ij}$. In spite of its simplicity this potential shows two characteristics

- ultraviolet (*UV*) divergence;
- infrared (*IR*) divergence.

The *UV* divergence corresponds to the unbound interaction for two closely approaching bodies, $\lim_{r_{ij}\to 0} U_{ij} = +\infty$, while the *IR* divergence is due to the long-range nature of gravity which never exactly vanishes at any distance ($F_{ij} > 0$, $\forall r_{ij}$). We will address and explore this topic in Chap. 6; here, I simply highlight some of the issues it raises in the context of the N-body problem.

First of all, the system of equations of motion is a differential system of order $6N$ which has only 10 classical first integrals and constants of motions, for any value of N. A theorem by Heinrich Bruns (1848-1919) shows that for $N \geq 3$ there are not

[4] Such a field can be considered as representing the gravitational force exerted by a spherical gaseous cloud where the N bodies are embedded in.

[5] Note that we are considering Q as defined in Eq. 2.55.

enough independent constants of motions to solve it explicitly. This means that only the 2-body problem can be reduced to quadratures and, so, has an explicit solution.

Apart from this issue, the system of equations of motion—which, as discussed above, must be approached numerically—is highly non-linear and potentially singular due to the ultraviolet (*UV*) divergence of the Newtonian potential. This has the important implication of making the system inherently *chaotic*. The chaoticity of the classical Newtonian N-body problem depends, indeed, on its non-linearity and manifests as an extreme sensitivity to perturbations. Two N-body systems starting from nearly identical initial conditions follow trajectories in the six-dimensional phase space that eventually diverge from each other. The rate of this divergence can be quantified by the so-called Lyapunov exponents: the larger (positive) the Lyapunov exponent, the faster the separation of the two trajectories. This behaviour corresponds to deterministic chaoticity.

Finally, every numerical attempt to solve the system faces the high computational cost of the evaluation of its right hand sides, i.e. the pair forces. Actually, due to the *IR* divergence of Newtonian force, this evaluation implies a cost of order N^2, because the number of distinct pairs is $N(N-1)/2$.

Because each (i, j) force evaluation requires computing a square root to obtain r_{ij}, the $O(N^2)$ scaling quickly becomes computationally prohibitive for large-N systems, even for fastest computers.

To synthesize:

- the ultraviolet (*UV*) divergence implies the need for an extremely small numerical time step whenever two bodies come very close to each other;
- the infrared (*IR*) divergence implies that, at each time step, computing the force on the entire N-body configuration becomes computationally overwhelming when N exceeds certain thresholds.

2.1.3.7 Collisional and Collisionless N-Body Systems

The dynamics of a system of N interacting bodies can be, at first order, represented as if the generic ith body is subjected to the force deriving from a regular and smooth contribution of a slowly varying potential, $\nabla U(\mathbf{r}; t)$, caused by the influence of other bodies and a more rapidly varying (fluctuating) component ($\mathbf{F}_f$) superposed. This latter component should account for the random (stochastic) close passage of another, kth, body to the ith. During this close passage at a mutual distance $\sim b$, whose time duration is the so-called *fly-by* time, $\tau_{fb} \propto b/v_{ik}$ where v_{ik} is the absolute value of their relative velocity, the $i-k$ inter-particle force overwhelms the overall force. In formulas

$$\ddot{\mathbf{r}}_i = \nabla U(\mathbf{r}_i; t) + \mathbf{F}_f. \tag{2.68}$$

Obviously, if the chance of close encounters is negligible, the perturbing, stochastic, term can be neglected. The question is: under what conditions collisions and encounters are relevant?

Through a series of considerations, one finds that the total number of bodies is crucial in determining the relevance of the time dependent perturbation over the slowly varying mean field. These considerations involve the concept of *collisionality* of a dynamical system.

2.1.3.8 Physical Collisions and Gravitational Encounters

Let us start with the discussion of the role of *physical* collisions. In this context the concept of the *mean free path* is relevant. It is defined as the average distance a particle travels before undergoing a *physical* collision with another particle. In the absence of any distance-dependent interaction potential among particles of the system, its standard expression, as derived from kinetic theory of gases, is

$$\lambda = \frac{1}{\sqrt{2}n\sigma}, \tag{2.69}$$

where n is the particle number density (number per unit volume) and σ is the *cross-sectional* area for the kind of collision considered.

Based on elementary considerations, the cross section for physical collisions between two of N randomly moving spherical bodies of equal radius r_* is given by $\sigma = \pi(2r_*)^2 = 4\pi r_*^2$. Assuming a uniform distribution of the N bodies within a spherical volume of radius R, it follows that:

$$\lambda = \frac{R^3}{3\sqrt{2}Nr_*^2} = \frac{R}{3\sqrt{2}N}\left(\frac{R}{r_*}\right)^2. \tag{2.70}$$

The above relation allows the expressions of the ratio of λ to the particle, r_*, and system, R, sizes as

$$\frac{\lambda}{r_*} = \frac{1}{3\sqrt{2}N}\left(\frac{R}{r_*}\right)^3, \qquad \frac{\lambda}{R} = \frac{1}{3\sqrt{2}N}\left(\frac{R}{r_*}\right)^2. \tag{2.71}$$

The more relevant of the two quantities above is surely λ/R, because $\lambda/(2R) < 1$ means that, in the average, a particle undergoes a physical collision before crossing the entire system where it moves in.

To make qualitative considerations on the N-body dynamics, another important reference scale is the average inter-collision time, $\tau = \lambda/\langle v\rangle$, where $\langle v\rangle$ is an estimate of the average absolute value of the relative velocity of the test object respect to the other $N-1$. A good estimate of $\langle v\rangle$ is the *virial* speed (Sect. 2.1.3.4) approximated by

$$v_{vir} = \sqrt{\frac{GM}{R}}, \tag{2.72}$$

Table 2.2 Order-of-magnitude estimates of collision parameters for various stellar systems, assuming uniform density and geometric cross sections (Eqs. 2.71, 2.73). Note: For galaxy clusters, r_* is the typical galactic radius; the actual τ time scales are shorter due to gravitational focusing and tidal interactions. As reference, the age of the Universe is 13.7 Gyr

System	N	m ($M_\odot$)	r_* (pc)	R (pc)	λ/r_*	λ/R	τ (Gyr)
Open cluster	10^2	1	10^{-8}	10	2.36×10^{24}	2.36×10^{15}	1.11×10^{15}
Globular cluster	10^6	1	10^{-8}	50	2.95×10^{22}	5.89×10^{12}	3.11×10^{11}
Galactic nucleus	10^6	1	10^{-8}	0.1	2.36×10^{14}	2.36×10^{7}	1.11×10^{2}
Galaxy cluster	10^3	10^{11}	10^4	3×10^6	6.36×10^{3}	21.2	5.20×10^{2}

where M is the total system mass. Assuming all the system objects to have the same mass m, the inter-collision time is

$$\tau = \frac{\lambda}{v_{vir}} = \frac{1}{3r_*^2\sqrt{2Gm}}\sqrt{\frac{R^7}{N^3}}. \tag{2.73}$$

Equations 2.71–2.73 indicate that poorly populated systems (low values of N) are less *collisional* than those with larger N (many body systems). This is because both λ and τ are decreasing functions of N at fixed values of R and m. In particular, the decrease of τ is steeper than that of λ: $\tau \propto N^{-3/2}$, while $\lambda \propto N^{-1}$.

The above considerations suggest that physical collisions are relevant in fluids, which are thus called *collision* dominated systems, but not in gravitational N-body system except for peculiar cases.

In Table 2.2 some characteristic values of parameters identifying some astronomical N-body systems are reported. In all the cases considered, except for the cases of galactic nuclei and clusters of galaxies, physical collisions are absolutely infrequent given the very large values of λ/R as well as of τ. In particular, the only regions of the universe where star-star runaway collisions may happen are the densest zones around a galactic centre.

Note that the cross section considered in Eq. 2.70 above is the so-called *geometrical* cross section, measuring the likelihood of actual physical collisions in absence of a distance-dependent interaction potential between the objects in consideration. When dealing with N massive bodies interacting via Newtonian gravity the above σ is underestimating the real one, σ_{eff}, which should account for the attractive characteristic of gravity. This phenomenon (gravitational *focusing*) leads to

$$\sigma_{eff} = \sigma(1+\alpha), \tag{2.74}$$

where $\alpha > 0$ for attractive forces ($\alpha < 0$ in the repulsive case) is an increasing function of the interacting bodies masses and a decreasing function of the relative speed and radius of the bodies. Elementary considerations based on conservation of energy and angular momentum in the centre-of-mass reference frame lead, upon the hypothesis of equal mass, m, and radius, r_*, for the interacting bodies, to

$\alpha = 2Gm/(r_* v_{vir}^2)$. When α is large enough, the trajectories of objects in the N-body system deviate from those they would follow in the smooth overall potential, U in Eq. 2.68, due to the sudden or cumulative effect of what are called *gravitational encounters*, represented by the fluctuating force $\mathbf{F}_f$. In this context we note that encounters may be of two kinds. In some cases, just few unfrequent but strong encounters (*close encounters*) may lead to a substantial alteration of the test object trajectory in the mean field, possibly resulting also in its acquisition of a high positive orbital energy, E, to be expelled from the cluster (*ejection* phenomenon). An analogous result of deviation from the unperturbed orbit can derive by a succession of many weak encounters, with a possible progressive increase of E over the binding threshold (*evaporation* phenomenon).

It will be seen in the following how this latter cumulative effect of weak encounters can be reasonably treated as a sequence of two-body interactions, which are by far more frequent than multi-body interactions because two-body interaction probability depends on the square of the average number density, n, of scatterers while an encounter among k objects has a probability $\propto n^k$ which, in a low dense environment, is much lower for larger k.

It is easily seen that in a Poissonian view of an N-body system enclosed in a volume, the mass density fluctuations are such that their ratio to the average density value, $\langle \rho \rangle$, is

$$S_\rho = \frac{\delta\rho}{\langle \rho \rangle} = \frac{\sqrt{N}}{N} = \frac{1}{\sqrt{N}}, \tag{2.75}$$

which is progressively decreasing at increasing N. This means that the limit of non-collisionality (*smooth field*) regime is that of large N. Fluctuations over a regular density distribution are instead relevant for small N, corresponding to collisional regime.[6]

The question is: how can the range of N for collisionality and non-collisionality be evaluated?

Of course, there is no precise and straightforward way to determine this range but only indicative estimates, as we now discuss.

Let us consider a homogeneous distribution of bodies, $\rho = const. = \rho_0$, within a sphere of radius R, whose total mass is M_t. The test particle at distance r from the centre of the sphere feels a gravitational radial acceleration towards the centre (see Sect. 2.2.4) of magnitude

$$a_r = G\frac{M(r)}{r^2} = G\frac{4\pi}{3}\rho_0 r. \tag{2.76}$$

Now, in an N-body system sampled from a uniform spherical distribution, consider a generic test particle i of mass m_i moving in the overall mean force field.

[6] Note that here the term “collision” is synonym of “gravitational encounter” and not of “physical collision”.

It can occasionally meet a background particle (the jth, of mass m_j) at a mutual distance r_{ij}. The acceleration induced by this passing object j on the test particle i is, in magnitude,

$$a_{ij} = G\frac{m_j}{r_{ij}^2}. \tag{2.77}$$

The ratio, ξ_{ij}, between the individual encounter acceleration (2.77) and the overall one (2.76) is

$$\xi_{ij} = \frac{m_j}{M(r)}\left(\frac{r}{r_{ij}}\right)^2 = \frac{1}{N}\frac{m_j}{\langle m \rangle}\frac{R^3}{r_{ij}^2 r}, \tag{2.78}$$

where $\langle m \rangle = M_t/N$ is the average mass of the system particles.

When $\xi_{ij} = 1$, the single i-j encounter has on the test ith object the same role of the general force field of the $N-2$ particles, so that we can define a sort of threshold distance $r_{ij,eff}$ as

$$r_{ij,eff} = \sqrt{\frac{m_j}{\langle m \rangle}\frac{R^3}{N}\frac{1}{r}}, \tag{2.79}$$

which, in the hypothesis of equal mass objects, $\langle m \rangle = m_j = m$, reduces to

$$r_{ij,eff} = \frac{R^{3/2}}{\sqrt{Nr}} = \frac{\sqrt{N}}{N}\frac{R^{3/2}}{\sqrt{r}}. \tag{2.80}$$

The definition above corresponds to the fact that encounters at distances $r_{ij} \leq r_{ij,eff}$ are inducing on the test particle an acceleration larger than that produced by the overall regular potential. Of course, particles travelling deeper in the distribution suffer more from close encounters than particles in the outskirts. Note in Eq. 2.80 that the coefficient $\sqrt{N}/N$ of $R^{3/2}/\sqrt{r}$ is exactly the S_ρ relative density fluctuation given in Eq. 2.75. The minimum value of $r_{ij,eff}$ is reached at the sphere surface, and is $R/\sqrt{N}$.

A good reference value for $r_{ij,eff}$ is its mass-averaged value

$$\langle r_{ij,eff} \rangle = \frac{6}{5}\frac{\sqrt{N}}{N}R. \tag{2.81}$$

The value above is significant when compared with the average closest neighbour distance that, in a uniform distribution in 3D, is (see Eq. 2.85 in Exercise 2.1)

$$\langle d_{cn} \rangle \simeq \frac{0.55}{\sqrt[3]{n}} \simeq 0.89\frac{R}{\sqrt[3]{N}}. \tag{2.82}$$

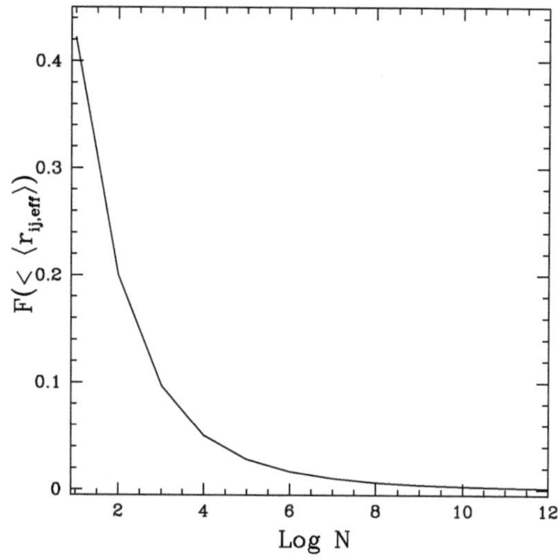

Fig. 2.2 The fraction of objects with a closest neighbour distance less than $\langle r_{eff} \rangle$, $F(< \langle r_{ij,eff} \rangle)$, as a function of the decimal logarithm of the total number of objects

To evaluate quantitatively the role of granularity over the mean field, let us estimate the likelihood of actual passages of objects close to another by means of the evaluation of the number of encounters, assuming for the closest neighbour distance a Gaussian distribution with mean $\langle d_{cn} \rangle$ and standard deviation $\sigma_{d_{cn}} = 0.42R/N^{1/3}$. Of course, granularity effects are important in the low tail of the distribution of this closest neighbour distance, which is quantitatively given by the cumulative distribution function (CDF), $F(< \langle r_{ij,eff} \rangle)$. In this regard, note that the normalized CDF of the Gaussian distribution of a parameter x truncated at $x < 0$ is

$$F(< x) = \frac{\mathrm{erf}\left(\frac{x - \langle d_{cn} \rangle}{\sqrt{2}\sigma_{d_{cn}}}\right) - \mathrm{erf}\left(-\frac{\langle d_{cn} \rangle}{\sqrt{2}\sigma_{d_{cn}}}\right)}{1 - \mathrm{erf}\left(-\frac{\langle d_{cn} \rangle}{\sqrt{2}\sigma_{d_{cn}}}\right)}. \tag{2.83}$$

The fraction of objects having a closest neighbour distance less or equal to $\langle d_{cn} \rangle$ is independent of N, and is a little less than 50%. On the other side, the fraction $F(< \langle r_{ij,eff} \rangle)$ is a decreasing function of N, as shown by Fig. 2.2, going from $\approx 42.2\%$ for $N = 10$ to $\approx 0.13\%$ for $N = 10^{12}$. The behaviour of the curve in the figure suggests a sufficiently clear division between collisional systems as those with $N \lesssim 10^4$ and collisionless as those with $N \gtrsim 10^7$. Systems in the $10^4 \div 10^7$ range of N are just partly collisional (in their denser regions).

Exercise 2.1 Evaluate the closest neighbour distance in a uniform distribution of N particles within a sphere of radius R. Compare to the average interparticle distance.

Solution Let us assume that the number of particles that lie within a distance D from a given (arbitrary) particle of the set follows a Poisson distribution. This means that the probability that the particle has no neighbours within distance D is

$$P_0(D) = \mathrm{e}^{-nV_D} = \mathrm{e}^{-\frac{4\pi}{3}D^3},$$

where n is the, constant, number density $n = 3N/(4\pi R^3)$, and V_D the volume of the sphere of radius D. Consequently, from elementary statistics, the probability density function of finding the first neighbour at exactly distance D is given by the derivative with respect to D of the *survival* function $S(D) = 1 - P_0(D)$, which is

$$P(D) \equiv \frac{\mathrm{d}S}{\mathrm{d}D} = 4\pi n D^2 \mathrm{e}^{-\frac{4\pi}{3}nD^3}. \tag{2.84}$$

Given the above relation, the closest neighbour average distance is

$$\langle D_{cn} \rangle = \int_0^\infty DP(D)\mathrm{d}D,$$

that by substitution from Eq. 2.84 leads to

$$\langle d_{cn} \rangle = 4\pi n \int_0^\infty D^3 \mathrm{e}^{-\frac{4\pi}{3}nD^3} \mathrm{d}D.$$

The change of variables $y = \sqrt[3]{(4\pi/3)}D$ transforms the integral above in the standard form

$$\langle d_{cn} \rangle = 3\sqrt[3]{\frac{3}{4\pi n}} \int_0^\infty y^3 \mathrm{e}^{-y^3} \mathrm{d}y = \frac{R}{\sqrt[3]{N}} \Gamma(4/3),$$

where $\Gamma(4/3)$ is the Euler Gamma function of argument 4/3, $\Gamma(4/3) \simeq 0.89298$, so that, finally

$$\langle D_{cn} \rangle \simeq 0.89298 \frac{R}{\sqrt[3]{N}}. \tag{2.85}$$

2.1.3.9 Two-Body Relaxation

As said above, systems not too populated are collision-dominated, in the sense that, if sufficiently old, gravitational encounters have been able to perturb substantially the orbits of particles with respect to those they would have in the general smooth potential. When this happens, we say that the system has reached a *relaxed* state, to which a further slower evolution of its overall properties follows.

The term *sufficiently old* requires an explanation, which can be given by the introduction of the phenomenon of two-body relaxation. As it was said in the previous subsection, the most frequent mutual gravitational encounters among objects in a Newtonian gravitating system of particles are those between two bodies. A study of the relaxation process can thus be done by considering two bodies as passing by each other at a distance sufficiently smaller than the average inter particle distance,[7] so that the overall potential from all the other particles can be considered negligible. Under this hypothesis, the effect of the pair interaction can be evaluated by the solution of the Newtonian two-body problem (see Appendix A).

The solution of the two-body problem is uniquely determined by their initial conditions that, in their turns, determine the values of the specific (per unit mass) mechanical energy, E, linear, $\mathbf{Q}$, and angular momentum, $\mathbf{L}$. It is known, indeed, that the trajectories are conic sections: *ellipses* for $E < 0$ (eccentricity $e < 1$, *parabolas*) for $E = 0$ ($e = 1$), hyperbolas for $E > 0$ ($e > 1$). Clearly, if $E < 0$ the trajectories of the two bodies are closed, with the two bodies revolving each other with periodicity. So, 2-body relaxation concerns only the case $E \geq 0$, the case of $E = 0$ being only a limiting boundary. Of course, the large the value of $E > 0$, the lesser the curvature of the hyperbola. In the limit $E \to +\infty$ the trajectories (in absence of external perturbation) are straight lines.

The *relative* motion is that studied in a reference frame centred in one of the two interacting bodies whose masses are m and M. The motion is planar, and letting $\mathbf{V} = \dot{\mathbf{r}}$ the relative velocity, it is

$$E = \frac{1}{2}V^2 - G\frac{m+M}{r}, \tag{2.86}$$

and

$$\mathbf{L} = \mathbf{r} \wedge \mathbf{V}. \tag{2.87}$$

This frame is, obviously, noninertial. If we denote by $\mathbf{v}_m$ and $\mathbf{v}_M$ the absolute velocities (those referred to an inertial frame, like the centre of mass frame) of the objects with mass m and M respectively, and we assume time t such that $t = 0$ at pericentre (minimum distance between the two bodies), we can define the total velocity variation upon the hyperbolic encounter as

$$\Delta\mathbf{v} \equiv \lim_{t\to+\infty} \mathbf{v}(t) - \lim_{t\to-\infty} \mathbf{v}(t), \tag{2.88}$$

and so its connection with the variation of the relative velocity, $\Delta\mathbf{V}$, is

$$\Delta\mathbf{v}_m - \Delta\mathbf{v}_M = \Delta\mathbf{V}. \tag{2.89}$$

[7] The average inter particle distance in a spherical and uniform distribution is $\langle d \rangle \simeq 1.61\langle d_{cn}\rangle$ (see Exercise 2.1).

In the inertial frame, the momentum is conserved and the velocity of the 2-body centre of mass

$$\mathbf{v}_G = \frac{m\mathbf{v}_m + M\mathbf{v}_M}{m + M} \tag{2.90}$$

is constant and is equal to zero in the centre of mass frame. Putting together Eq. 2.89 and Eq. 2.90 gives

$$\begin{cases} \Delta\mathbf{v}_m = \dfrac{M}{m+M}\Delta\mathbf{V}, \\ \Delta\mathbf{v}_M = -\dfrac{m}{m+M}\Delta\mathbf{V}. \end{cases} \tag{2.91}$$

It is now needed to evaluate Delta**V**, using the 2-body problem solution in the relative frame.

As it is seen in Appendix A, using standard (r, θ) polar coordinates in the xy-plane of motion, the trajectory of either the m or M body around the other is obtained as the conic

$$r(\theta) = \frac{p}{1 + e\cos(\theta - \theta_0)}, \tag{2.92}$$

where θ_0 is the position angle at pericentre passage, $p = L^2/[G(m + M)]$ is the semi-latus rectum of the conic (often called the parameter), and $e = \sqrt{1 + 2EL^2/[G(m + M)]^2}$ its eccentricity.

If $E > 0$, the particle is travelling along a hyperbola starting from infinite distance from the other with speed V_∞. Defining (see Fig. 2.3) $\psi = \pi - \theta$ the expression above conserves its form

$$r(\theta) = \frac{p}{1 + e\cos(\psi - \psi_0)}, \tag{2.93}$$

to give

$$\dot{r} = \frac{e}{p}r^2\dot{\psi}\sin(\psi - \psi_0) = \frac{e}{p}L\sin(\psi - \psi_0) = \frac{e}{p}V_\infty b\sin(\psi - \psi_0), \tag{2.94}$$

having taken into account that $r^2\dot{\psi} = L$ and $L = V_\infty b$, with b impact parameter.

Having chosen counting time t such that $t = 0$ at pericentre, the condition $\lim_{t\to-\infty} r = +\infty$ implies $\lim_{t\to-\infty} \psi = 0$, and Eq. 2.94 evaluated at $t \to -\infty$ gives

$$-V_\infty = \frac{e}{p}V_\infty b\sin(-\psi_0), \tag{2.95}$$

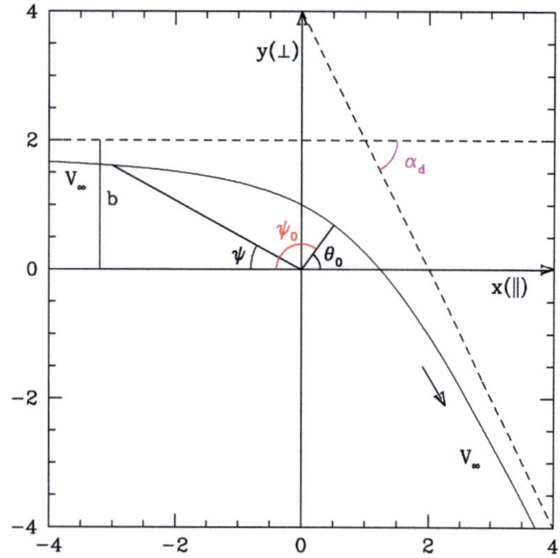

Fig. 2.3 Trajectory of a body coming from infinite distance with speed V_∞ to the closest approach (at θ_0) to another body placed at the origin of the coordinate system

that is

$$\sin\psi_0 = \frac{p}{eb}. \tag{2.96}$$

At the same time, evaluating 2.93 at $t \to -\infty$ we have

$$\cos\psi_0 = -\frac{1}{e}, \tag{2.97}$$

that together with the previous expression gives eventually to

$$\tan\psi_0 = -\frac{p}{b} = -\frac{bV_\infty^2}{G(m+M)}. \tag{2.98}$$

The deflection angle α_d (see Fig. 2.3) is the angle between the two asymptotes of the hyperbolic trajectory which is such that (by symmetry) $\alpha_d + \pi = 2\psi_0$.

The knowledge of $\alpha_d = 2\psi_0 - \pi$ allows the evaluation of the variation from $t \to -\infty$ to $t \to +\infty$ of the components of the relative velocity $\mathbf{V}$ along the x-axis *parallel* ($\|$) to $\mathbf{V}_\infty$ and along the y-axis *perpendicular* ($\perp$), to it[8]

$$\begin{aligned}\Delta\mathbf{V}_\| &= V\infty\cos\alpha_d - V_\infty = V_\infty(\cos\alpha_d - 1) = -V_\infty(1+\cos 2\psi_0) = \\ &= -\frac{2V_\infty}{1+\tan^2\psi_0} = -2V_\infty\left[1+\frac{b^2V_\infty^4}{G^2(m+M)^2}\right]^{-1},\end{aligned} \tag{2.99}$$

[8] Remember the identities $\cos 2x = 2\cos^2 x - 1$ and $\sin 2x = 2\sin x\cos x$.

$$\Delta \mathbf{V}_\perp = -V_\infty \sin\alpha_d = -V_\infty \sin 2\psi_0 = -2V_\infty \sin\psi_0 \cos\psi_0 = \\ = -\frac{2V_\infty \tan\psi_0}{1+\tan^2\psi_0} = \frac{2bV_\infty^3}{G(m+M)}\left[1+\frac{b^2V_\infty^4}{G^2(m+M)^2}\right]^{-1}. \tag{2.100}$$

Recalling Eq. 2.91, the absolute velocity variations are

$$\begin{cases} \Delta \mathbf{v}_{m\parallel} = \dfrac{M}{m+M}\Delta \mathbf{V}_\parallel = -\dfrac{2MV_\infty}{m+M}\left[1+\dfrac{b^2V_\infty^4}{G^2(m+M)^2}\right]^{-1}, \\ \Delta \mathbf{v}_{m\perp} = \dfrac{M}{m+M}\Delta \mathbf{V}_\perp = \dfrac{2MbV_\infty^3}{G(m+M)^2}\left[1+\dfrac{b^2V_\infty^4}{G^2(m+M)^2}\right]^{-1}, \end{cases} \tag{2.101}$$

and (Eq. 2.90) $\Delta \mathbf{v}_{M\parallel} = -(m/M)\Delta \mathbf{v}_{m\parallel}$ and $\Delta \mathbf{v}_{M\perp} = -(m/M)\Delta \mathbf{v}_{m\perp}$.

Note that the quantities above can be expressed also in terms of $E = V_\infty^2/2$ as

$$\begin{cases} \Delta \mathbf{v}_{m\parallel} = -\dfrac{2M\sqrt{2E}}{m+M}\left[1+\dfrac{b^2 4E^2}{G^2(m+M)^2}\right]^{-1}, \\ \Delta \mathbf{v}_{m\perp} = \dfrac{2Mb(2E)^{3/2}}{G(m+M)^2}\left[1+\dfrac{b^2 4E^2}{G^2(m+M)^2}\right]^{-1}, \end{cases} \tag{2.102}$$

that in the regime of $E \gg 0$ (weak scattering regime) approximate to

$$\begin{cases} \Delta \mathbf{v}_{m\parallel} \simeq -\dfrac{1}{2}\dfrac{G^2M(m+M)}{b^2}\sqrt{\dfrac{2}{E^3}}, \\ \Delta \mathbf{v}_{m\perp} \simeq \dfrac{GM}{b}\sqrt{\dfrac{2}{E}}, \end{cases} \tag{2.103}$$

equivalent to

$$\begin{cases} \Delta \mathbf{v}_{m\parallel} \simeq -\dfrac{2G^2M(m+M)}{b^2V_\infty^3}, \\ \Delta \mathbf{v}_{m\perp} \simeq \dfrac{2GM}{bV_\infty}. \end{cases} \tag{2.104}$$

It is easily seen (see Exercise 2.2) that, in the hypothesis of uniform distribution of the deflection angle α_d, the average $\langle \Delta \mathbf{V}_\perp \rangle$ is zero over every interval $[-a, a]$ of α_d while the average $\langle \Delta \mathbf{V}_\parallel \rangle$ is nonzero and equal to $a + \sin a \cos a$.

On the other side, the variances $\langle (\Delta \mathbf{V}_\parallel - \langle \Delta \mathbf{V}_\parallel \rangle)^2 \rangle_{\alpha_d}$ and $\langle (\Delta \mathbf{V}_\perp - \langle \Delta \mathbf{V}_\perp \rangle)^2 \rangle_{\alpha_d}$ are not zero (see again Exercise 2.2).

We can estimate the *two-body relaxation time* as follows. Consider *test* objects of mass M (also called *bullets*) moving through an environment of *field* objects of mass m (or targets). These test objects come from $t \to -\infty$ with speed V_∞ along a given direction. The relaxation time is defined as the time needed for them to transfer a certain fraction of their kinetic energy from the initial direction (along $\mathbf{v}_{M\parallel}$) to the perpendicular one ($\mathbf{v}_{M\perp}$). Let us proceed to obtain an evaluation of this relaxation time.

The differential number of scatterings per unit volume suffered by the test object in the interval $\mathrm{d}t$ with impact parameter in the interval $[b, b + \mathrm{d}b]$ is $\mathrm{d}n_s = v_M \,\mathrm{d}t\, 2\pi b\, \mathrm{d}b\, f(\mathbf{v}_m)\mathrm{d}^3\mathbf{v}_m$, where $f(\mathbf{v}_m)\,\mathrm{d}^3\mathbf{v}_m$ is the differential number per unit volume of field objects with velocity in the interval $[\mathbf{v}_m, \mathbf{v}_m + \mathrm{d}^3\mathbf{v}_m]$—a quantity assumed here to be independent of space and time.[9] Consequently, the variance of $\Delta\mathbf{v}_{M\perp}$ at the generic time t is

$$\langle \Delta\mathbf{v}^2_{M\perp}\rangle_t = \int_0^\infty \left[\frac{2mbV^3_\infty}{G(m+M)^2}\right]^2 \frac{2\pi b\,\mathrm{d}b}{\left[1+\frac{b^2V^4_\infty}{G^2(m+M)^2}\right]^2} \int_{-\infty}^{t} \mathrm{d}t \iiint_{\mathbb{R}^3} v_M f(\mathbf{v}_m)\mathrm{d}^3\mathbf{v}_m. \tag{2.105}$$

Letting $x = (b^2V^4_\infty)/[G^2(m+M)^2]$ transforms the integral over b into

$$\frac{4\pi G^2m^2}{V^2_\infty}\int_0^\infty \frac{x}{(1+x)^2}\mathrm{d}x. \tag{2.106}$$

The above integral over x is easily solved adding and subtracting 1 in the numerator, leading to

$$\int \frac{x}{(1+x)^2}\mathrm{d}x = \ln(1+x) + \frac{1}{1+x}, \tag{2.107}$$

which converges in $x = 0$, while it does not for $x \to +\infty$. This problem is overcome because the extremes of integration are naturally defined by the two conditions of: (*i*) spatial finiteness of the system of field bodies, inducing an upper boundary x_{max} in the integrand and (*ii*) the condition of hyperbolic pair relative motion, which sets a minimum, x_{min}. The value of x_{max} is set by adopting a maximum impact parameter b_{max}, reasonably corresponding to the size of the system, giving

$$x_{max} = \frac{b^2_{max}V^4_\infty}{G^2(m+M)^2}. \tag{2.108}$$

By means of energy conservation, the value of the minimum impact parameter is

$$b_{min} = \frac{G(m+M)}{\frac{1}{2}\left(V^2_p - V^2_\infty\right)}, \tag{2.109}$$

[9] The function $f(\mathbf{v}_m)$ is called velocity distribution function.

where V_p is the velocity at pericentre, so that

$$x_{min} = \frac{1}{\left[\frac{1}{2}\left(V_p^2/V_\infty^2 - 1\right)\right]^2}. \tag{2.110}$$

Consequently, it is

$$\langle \Delta \mathbf{v}_{M\perp}^2 \rangle_t = \frac{4\pi G^2 m^2}{V_\infty^2} v_M \Delta t\, n \left(\ln \frac{1+x_{max}}{1+x_{min}} + \frac{1}{1+x_{max}} - \frac{1}{1+x_{min}} \right), \tag{2.111}$$

where n is the number density of the field objects and Δt the time elapsed since the entrance of the test object in the scatterer distribution. In the limit $x_{max} \gg 1$ and $x_{min} \ll 1$ and assuming for V_∞ and v_M the same averaged value as the virial speed, v_{vir}, of the system, and n as constant, the 2-body *relaxation time* is defined as the interval of time Δt needed to reach a value of $\langle \Delta \mathbf{v}_{M\perp}^2 \rangle_t = \beta v_{vir}^2$, that is

$$t_{rel} = \frac{\beta v_{vir}^3}{8\pi G^2 n\, m^2 \ln \dfrac{b_{max} v_{vir}^2}{G(m+M)}}. \tag{2.112}$$

The above expression can be also written introducing at denominator the mass density of field stars $\rho = nm$; the logarithm at denominator is often called Coulomb logarithm, $\ln \Lambda$. The assumption $\beta = 1$ leads to an expression of the relaxation time that, using proper astrophysical unit of measures, is

$$t_{rel} = \frac{2.1 \times 10^9 v_{vir}^3}{(m/M_\odot)^2\, n \ln \Lambda} \text{ yr}, \tag{2.113}$$

with v_{vir} in km s^{-1} and n in pc^{-3}. It is remarked the inverse square dependence of t_{rel} on the mass of field stars. The value of $\ln \Lambda$ is quite uncertain, by the way its role on the determination of t_{rel} is not too strong. Actually, an uncertainty ϵ_Λ in Λ induces a relative uncertainty in t_{rel}

$$\left| \frac{\epsilon_{t_{rel}}}{t_{rel}} \right| = \frac{1}{t_{rel}} \left| \frac{\partial t_{rel}}{\partial \Lambda} \right| |\epsilon_\Lambda| = \frac{1}{\ln \Lambda} \left| \frac{\epsilon_\Lambda}{\Lambda} \right|, \tag{2.114}$$

so that the relative error in Λ is quenched by an order of magnitude in the corresponding relative error in t_{rel}.

Exercise 2.2 Calculate the average, over deflection angle α_d, of $\Delta \mathbf{V}_\parallel$ and $\Delta \mathbf{V}_\perp$ assuming a uniform distribution of α_d in the whole range of its variation.

Solution Assuming α_d uniformly distributed over the generic interval $[-a, a]$ with $a > 0$, the two averages, $\langle \Delta \mathbf{V}_\parallel \rangle_{\alpha_d}$ and $\langle \Delta \mathbf{V}_\perp \rangle_{\alpha_d}$, require the computation of the integrals

$$\langle \Delta \mathbf{V}_\parallel \rangle_{\alpha_d} = -\frac{V_\infty}{a} \int_{-a}^{a} \frac{\mathrm{d}\alpha_d}{1 + \tan^2 \psi_0} = -\frac{2V_\infty}{a} \int_{\psi_0(-a)}^{\psi_0(a)} \frac{\mathrm{d}\psi_0}{1 + \tan^2 \psi_0}, \tag{2.115}$$

and

$$\begin{aligned} \langle \Delta \mathbf{V}_\perp \rangle_{\alpha_d} &= -\frac{V_\infty}{a} \int_{-a}^{a} \frac{\tan \psi_0}{1 + \tan^2 \psi_0} \mathrm{d}\alpha_d = \\ &= -\frac{2V_\infty}{a} \int_{\psi_0(-a)}^{\psi_0(a)} \frac{\tan \psi_0}{1 + \tan^2 \psi_0} \mathrm{d}\psi_0. \end{aligned} \tag{2.116}$$

Given that $\psi_0 = (\alpha_d + \pi)/2$, the limits of integrations in the integrals above are $\psi_0(-a) = (\pi - a)/2$ and $\psi_0(a) = (\pi + a)/2$. Note that the constraint $E > 0$ implies an upper limit, a_{max}, to a, that results to be

$$a_{max} = \arctan \frac{b_{min} V_\infty^2}{G(m + M)},$$

where b_{min} has been defined in Eq. 2.109.

To solve the integral in Eq. 2.115 we use the trigonometric identity $1 + \tan^2 \psi_0 = \sec^2 \psi_0$ to write the integral as

$$I_\parallel = \int \frac{\mathrm{d}\psi_0}{1 + \tan^2 \psi_0} = \int \cos^2 \psi_0 \, \mathrm{d}\psi_0.$$

This integral is easy solved making use of the cosine double-angle formula cited in the footnote 8, to have

$$I_\parallel = \int \cos^2 \psi_0 \, \mathrm{d}\psi_0 = \frac{1}{2} \int (1 + \cos 2\psi_0) \, \mathrm{d}\psi_0 = \frac{1}{2} \left(\psi_0 + \frac{\sin 2\psi_0}{2} \right) + c.$$

Evaluating the above in the limits of integration $(\pi - a)/2$ and $(\pi + a)/2$ we have

$$\langle \Delta \mathbf{V}_\parallel \rangle_{\alpha_d} = -\frac{V_\infty}{a} (a - \sin a). \tag{2.117}$$

Note that $\langle \Delta \mathbf{V}_\parallel \rangle_{\alpha_d} < 0 \quad \forall a \in]0, \pi/2]$.

Regarding to $\langle\Delta\mathbf{V}_{\perp}\rangle_{\alpha_d}$, the integral in Eq. 2.116 can be written as

$$I_{\perp} = \int \frac{\tan\psi_0}{1+\tan^2\psi_0}\mathrm{d}\psi_0 = \int \sin\psi_0\cos\psi_0\,\mathrm{d}\psi_0, = -\frac{1}{2}\cos^2\psi_0 + c,$$

that in the limits of integration $(\pi - a)/2$ and $(\pi + a)/2$ gives

$$\langle\Delta\mathbf{V}_{\perp}\rangle_{\alpha_d} = 0, \quad \forall a. \tag{2.118}$$

Let us now compute the variances $\langle\left(\Delta\mathbf{V}_{\parallel} - \langle\Delta\mathbf{V}_{\parallel}\rangle\right)^2\rangle_{\alpha_d}$ and $\langle(\Delta\mathbf{V}_{\perp} - \langle\Delta\mathbf{V}_{\perp}\rangle)^2\rangle_{\alpha_d} = \langle(\Delta\mathbf{V}_{\perp})^2\rangle_{\alpha_d}$. Accounting for Eqs. 2.99, 2.100, 2.117, and 2.118, the results are

$$\langle\left(\Delta\mathbf{V}_{\parallel} - \langle\Delta\mathbf{V}_{\parallel}\rangle\right)^2\rangle_{\alpha_d} = \frac{V_\infty^2}{a}\int_{(\pi-a)/2}^{(\pi+a)/2}\left(-\frac{2}{1+\tan^2\psi_0} + \frac{a-\sin a}{a}\right)^2\mathrm{d}\psi_0 =$$
$$= \frac{1}{2}V_\infty^2\left(1 - 2\frac{\sin^2 a}{a^2} + \frac{\sin a\cos a}{a}\right),$$

and

$$\langle(\Delta\mathbf{V}_{\perp})^2\rangle_{\alpha_d} = 4\frac{V_\infty^2}{a}\int_{(\pi-a)/2}^{(\pi+a)/2}\left(\frac{\tan\psi_0}{1+\tan^2\psi_0}\right)^2\mathrm{d}\psi_0 = \frac{1}{2}V_\infty^2\left(1 - \frac{\sin a\cos a}{a}\right).$$

Note that both are increasing function of a in $a \le \pi/2$ and are vanishing for $a \to 0$, while $\langle\left(\Delta\mathbf{V}_{\parallel} - \langle\Delta\mathbf{V}_{\parallel}\rangle\right)^2\rangle_{\alpha_d} \simeq 0.18943(V_\infty^2/2)$ and $\langle\Delta\mathbf{V}_{\perp}^2\rangle_{\alpha_d} = V_\infty^2/2$ for $a = \pi/2$.

The series expansion around $a = 0$ (small deflections, high values of E) are

$$\langle\left(\Delta\mathbf{V}_{\parallel} - \langle\Delta\mathbf{V}_{\parallel}\rangle\right)^2\rangle_{\alpha_d} = \frac{1}{2}V_\infty^2\left(\frac{2}{45}a^4 - \frac{2}{315}a^6 + O(a^8)\right),$$

and

$$\langle(\Delta\mathbf{V}_{\perp})^2\rangle_{\alpha_d} = \frac{1}{2}V_\infty^2\left(\frac{2}{3}a^2 - \frac{2}{15}a^4 + \frac{4}{315}a^6 + O(a^8)\right).$$

2.1.3.10 Dynamical Friction

As we saw from Exercise 2.2, the average of $\Delta\mathbf{V}_{\parallel}$ is nonzero and negative in intervals of α_d of interest, contrary to $\langle\Delta\mathbf{V}_{\perp}\rangle$. This has a consequence when making general considerations on the motion of test objects of mass M in a field of objects of mass m. Because the individual gravitational scattering induces a net reduction of the velocity parallel to the initial velocity, to evaluate the cumulative effect of weak scatterings along the test particle movement we can replace $\mathbf{V}_\infty$ with $\delta\mathbf{v} \equiv \mathbf{v}_m - \mathbf{v}_M$ such that, proceeding in a way similar to that for the computation of $\langle \Delta\mathbf{v}^2_{M\perp} \rangle$, we obtain a cumulative rate of encounters inducing an average acceleration in the time interval Δt that can be written as

$$\frac{\Delta\mathbf{v}_M^{(c)}}{\Delta t} = \int_0^\infty f(\mathbf{v}_m)\delta\mathbf{v}\, \mathrm{d}^3\mathbf{v}_m \int_0^\infty \frac{2m\|\delta\mathbf{v}\|}{m+M}\left[1+\frac{b^2\|\delta\mathbf{v}\|^4}{G^2(m+M)^2}\right]^{-1} 2\pi b\, \mathrm{d}b. \tag{2.119}$$

Letting

$$I_b = \int_0^\infty \left[1+\frac{b^2\|\delta\mathbf{v}\|^4}{G^2(m+M)^2}\right]^{-1} 2\pi b\, \mathrm{d}b, \tag{2.120}$$

and $\Gamma = \|\delta\mathbf{v}\|^2/[G(m+M)]$, it results

$$I_b = \frac{\pi}{\Gamma^2}\int_1^\infty \frac{\mathrm{d}(1+\Gamma^2 b^2)}{1+\Gamma^2 b^2} = \frac{\pi}{\Gamma^2}\int_1^\infty \frac{\mathrm{d}\Lambda}{\Lambda}, \tag{2.121}$$

upon the substitution $\Lambda = 1 + \Gamma^2 b^2$. The integral (2.121) is logarithmically diverging at the upper limit. Due to previous considerations (Sect. 2.1.3.9) the actual lower and upper integration limit should be $\Lambda_{min} = 1 + \Gamma^2 b^2_{min}$ and $\Lambda_{max} = 1 + \Gamma^2 b^2_{max}$, respectively. Thus, Eq. 2.119, becomes

$$\frac{\Delta\mathbf{v}_M^{(c)}}{\Delta t} = 2\pi G^2 m(m+M)\ln\frac{\Lambda_{max}}{\Lambda_{min}}\int_0^\infty f(\mathbf{v}_m)\frac{\mathbf{v}_m - \mathbf{v}_M}{\|\mathbf{v}_m - \mathbf{v}_M\|^3}\,\mathrm{d}^3\mathbf{v}_m. \tag{2.122}$$

In the hypothesis of isotropic velocity distribution for field objects, i.e. $f(\mathbf{v}_m) = f(v_m)$, the above expression reduces to

$$\frac{\Delta\mathbf{v}_M^{(c)}}{\Delta t} = -8\pi^2 G^2 m(m+M)\ln\frac{\Lambda_{max}}{\Lambda_{min}}\int_0^{v_M} f(v_m)v_m^2\,\mathrm{d}v_m\,\frac{\mathbf{v}_M}{v_M^3}. \tag{2.123}$$

This formula represents a *drag* deceleration, commonly referred to as *dynamical friction*—a phenomenon first described by S. Chandrasekhar in 1943 in the context of stellar dynamics.

Note that for small values of v_M the expression above can be well approximated by

$$\frac{\Delta \mathbf{v}_M^{(c)}}{\Delta t} = -\frac{8\pi^2}{3} G^2 m(m+M) \ln \frac{\Lambda_{max}}{\Lambda_{min}} f(0)\, \mathbf{v}_M, \tag{2.124}$$

which resembles the well-known Stokes law for the drag suffered by a body moving in a viscous fluid in the laminar regime. At the opposite extreme, if the integral in Eq. 2.123 is convergent in the upper limit v_M, the frictional force decays as v_M^{-2} at large v_M. Another relevant feature of dynamical friction deceleration is that for massive objects ($M \gg m$) it is proportional to M. This corresponds to the *gravitational focusing* of the field objects trajectories caused by the moving object of mass M, which corresponds to an overdensity at its back respect to its front, leading to a net deceleration (called *wake effect*).

2.2 Regular Distributions of Matter

Let $\mathbf{F}(\mathbf{r})$ be the gravity force exerted on the unitary mass in the point $\mathbf{r}$ by a regular distribution of matter with volume density $\rho(\mathbf{r})$ defined in a bound domain C (see Fig. 2.4). The generalization to a continuous distribution of the expression by sum of the total force exerted by a set of N point masses on a unit test mass located at $\mathbf{r}$ leads, straightforwardly, to the leads

$$\mathbf{F}(\mathbf{r}) = G \iiint_C \frac{\mathbf{r}' - \mathbf{r}}{\|\mathbf{r}' - \mathbf{r}\|^3} \rho(\mathbf{r}')\, \mathrm{d}^3\mathbf{r}'. \tag{2.125}$$

Clearly, the potential of this force is

$$U(\mathbf{r}) = G \iiint_C \frac{\rho(\mathbf{r}')}{\|\mathbf{r}' - \mathbf{r}\|} \mathrm{d}^3\mathbf{r}', \tag{2.126}$$

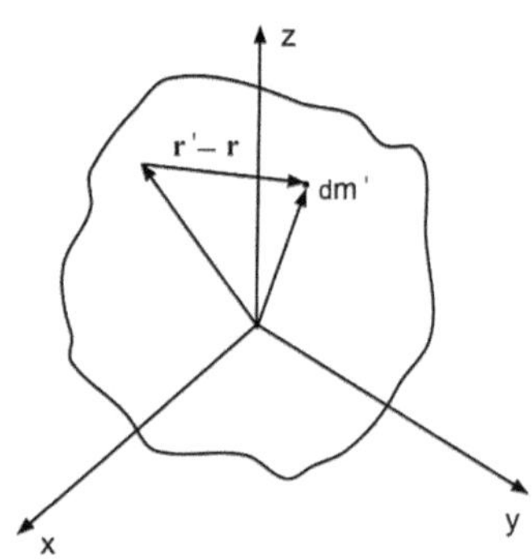

Fig. 2.4 The vector $\mathbf{r}$ indicates the position of the particle feeling the infinitesimal force generated by the particle of mass $\mathrm{d}m'$ placed in $\mathbf{r}'$

being

$$\nabla U = \mathbf{F}(\mathbf{r}). \tag{2.127}$$

2.2.1 *Poisson's and Laplace's Equations*

Let us see some characteristics of the force field (2.125) and its potential (2.126).

For this purpose, we distinguish the *internal* case, $\mathbf{r} \in C$, from the *external*, $\mathbf{r} \notin C$.

We have

$$\begin{aligned}\nabla \cdot \mathbf{F}(\mathbf{r}) &= G \iiint\limits_C \nabla \cdot \frac{\mathbf{r}' - \mathbf{r}}{\|\mathbf{r}' - \mathbf{r}\|^3} \rho(\mathbf{r}')\, \mathrm{d}^3\mathbf{r}' = \\ &= G \iiint\limits_C \left[\frac{1}{\|\mathbf{r}' - \mathbf{r}\|^3} \nabla \cdot (\mathbf{r}' - \mathbf{r}) + (\mathbf{r}' - \mathbf{r}) \cdot \nabla \frac{1}{\|\mathbf{r}' - \mathbf{r}\|^3} \right] \rho(\mathbf{r}')\, \mathrm{d}^3\mathbf{r}'.\end{aligned} \tag{2.128}$$

For every $\mathbf{r}$ and $\mathbf{r}'$ in $\mathbb{R}^3$, $\nabla \cdot (\mathbf{r}' - \mathbf{r}) = -3$, and, so

$$\nabla \frac{1}{\|\mathbf{r} - \mathbf{r}'\|^3} = -\frac{3}{\|\mathbf{r} - \mathbf{r}'\|^4} \nabla \|\mathbf{r} - \mathbf{r}'\| = \frac{3}{\|\mathbf{r} - \mathbf{r}'\|^4} \frac{\mathbf{r}' - \mathbf{r}}{\|\mathbf{r}' - \mathbf{r}\|}, \tag{2.129}$$

consequently

$$\nabla \cdot \mathbf{F}(\mathbf{r}) = G \iiint\limits_C \left[-\frac{3}{\|\mathbf{r}' - \mathbf{r}\|^3} + \frac{3(\mathbf{r}' - \mathbf{r}) \cdot (\mathbf{r}' - \mathbf{r})}{\|\mathbf{r}' - \mathbf{r}\|^5} \right] \rho(\mathbf{r}')\, \mathrm{d}^3\mathbf{r}'. \tag{2.130}$$

Now, if we evaluate the divergence of the force in an external point, $\mathbf{r} \notin C$, it is $\mathbf{r} \neq \mathbf{r}'$, consequently the integral in Eq. 2.130 is a proper integral whose integrand is zero. As a conclusion

$$\nabla \cdot \mathbf{F} = 0, \tag{2.131}$$

in the *empty* space around a distribution of matter. This means that the force field is *solenoidal* in the empty space; consequently, the *Laplace's* equation for $U(\mathbf{r})$ holds

$$\nabla \cdot \nabla U \equiv \nabla^2 U = 0, \tag{2.132}$$

holds. In other words, the field equation for the classical gravitational force in the empty space is just the Laplace's equation.

On the other hand, the evaluation of the force in a point within the matter distribution implies that the integral (2.130) is improper, the integrand being singular in $\mathbf{r}' = \mathbf{r}$. We can evaluate this improper integral as

$$\begin{aligned}\nabla \cdot \mathbf{F} &= G \lim_{\mathbf{r}' \to \mathbf{r}} \iiint_C \rho(\mathbf{r}') \nabla \cdot \frac{\mathbf{r}' - \mathbf{r}}{\|\mathbf{r}' - \mathbf{r}\|^3}\, \mathrm{d}^3\mathbf{r}' = \\ &= G \lim_{h \to 0} \iiint_{S_h(\mathbf{r})} \rho(\mathbf{r}') \nabla \cdot \frac{\mathbf{r}' - \mathbf{r}}{\|\mathbf{r}' - \mathbf{r}\|^3}\, \mathrm{d}^3\mathbf{r}',\end{aligned} \tag{2.133}$$

where $S_h(\mathbf{r}) = \left\{\mathbf{r}' : \|\mathbf{r}' - \mathbf{r}\| \le h\right\}$ is the spherical domain of radius h centred in $\mathbf{r}$. The integral above is evaluated by applying the mean theorem for integrals, first, and, after, the divergence theorem. The mean theorem allows taking out of the integral the density, $\rho(\mathbf{r}')$, as evaluated in a (unknown) point $\mathbf{r}_h$ within $S_h(\mathbf{r})$, while the divergence theorem can be used after we swapped $\nabla_{\mathbf{r}}$ with $-\nabla_{\mathbf{r}'}$,[10] thanks also to the continuity of $\rho(\mathbf{r}')$.

Finally

$$\begin{aligned}\nabla \cdot \mathbf{F} &= -G \lim_{h \to 0} \rho(\mathbf{r}_h) \iiint_{S_h(\mathbf{r})} \nabla_{\mathbf{r}'} \cdot \frac{\mathbf{r}' - \mathbf{r}}{\|\mathbf{r}' - \mathbf{r}\|^3} \mathrm{d}^3\mathbf{r}' = \\ &= -G \lim_{h \to 0} \rho(\mathbf{r}_h) \int_{\partial S_h(\mathbf{r})} \frac{\mathbf{r}' - \mathbf{r}}{\|\mathbf{r}' - \mathbf{r}\|^3} \cdot \mathbf{n}\, \mathrm{d}\sigma,\end{aligned} \tag{2.134}$$

where $\partial S_h(\mathbf{r})$ is the surface of the sphere $(\mathbf{r}' : \|\mathbf{r}' - \mathbf{r}\| = h)$ and $\mathbf{n}$ is the normal unit vector in the generic point of the sphere surface. Obviously, $\mathbf{n}$ points in the direction of $\mathbf{r}' - \mathbf{r}$, i.e.

$$\mathbf{n} = \frac{\mathbf{r}' - \mathbf{r}}{\|\mathbf{r}' - \mathbf{r}\|}, \tag{2.135}$$

so that

$$\begin{aligned}\nabla \cdot \mathbf{F} &= -G \lim_{h \to 0} \rho(\mathbf{r}_h) \int_{\partial S_h(\mathbf{r})} \frac{1}{\|\mathbf{r}' - \mathbf{r}\|^2}\, \mathrm{d}\sigma' = -G \lim_{h \to 0} \frac{\rho(\mathbf{r}_h)}{h^2} \int_{\partial S_h(\mathbf{r})} \mathrm{d}\sigma' = \\ &= -G \lim_{h \to 0} \frac{\rho(\mathbf{r}_h)}{h^2} 4\pi h^2 = -4\pi G \rho(\mathbf{r}),\end{aligned} \tag{2.136}$$

---

[10] The exchange of ∇r with $-\nabla r'$ is not a general property; in this case it is allowed because ∇ is applied to $(\mathbf{r}' - \mathbf{r})/\|\mathbf{r}' - \mathbf{r}\|$.

the final passage being justified by the continuity of the density $\rho(\mathbf{r})$. The final result corresponds to the so-called *Poisson's equation*

$$\nabla \cdot \mathbf{F} = \nabla \cdot \nabla U = \nabla^2 U = -4\pi G \rho(\mathbf{r}), \tag{2.137}$$

which is the gravitational field equation within the matter domain.

To resume, the field equations linking a regular matter density distribution $\rho(\mathbf{r})$ nonzero in C and zero outwards with the Newtonian potential $U(\mathbf{r})$ it generates are

$$\begin{cases} \nabla^2 U = 0, & \text{if } \mathbf{r} \notin C \text{ (Laplace's eq.)}, \\ \nabla^2 U = -4\pi G \rho(\mathbf{r}), & \text{if } \mathbf{r} \in C \text{ (Poisson's eq.)}. \end{cases} \tag{2.138}$$

Note 11

The definition of the force per unit mass via the integral 2.125 is valid everywhere. Actually, the integral is improper just for $\mathbf{r} = \mathbf{r}'$, but it always converges. Indeed, it results

$$0 \leq \left| \lim_{h\to 0} G \iiint\limits_{S_h(\mathbf{r})} \frac{\rho(\mathbf{r}')}{\|\mathbf{r}' - \mathbf{r}\|^3} (\mathbf{r}' - \mathbf{r})\, \mathrm{d}^3\mathbf{r}' \right| \leq G \lim_{h\to 0} \rho(\mathbf{r}_h) \iiint\limits_{S_h(\mathbf{r})} \frac{\mathrm{d}^3\mathbf{r}'}{\|\mathbf{r}' - \mathbf{r}\|^2}$$

$$= G \lim_{h\to 0} \rho(\mathbf{r}_h) \int_0^h 4\pi \frac{\cancel{\|\mathbf{r}' - \mathbf{r}\|^2}}{\cancel{\|\mathbf{r}' - \mathbf{r}\|^2}}\, \mathrm{d}\, \|\mathbf{r}' - \mathbf{r}\| = 4\pi G \rho(\mathbf{r}_h) \lim_{h\to 0} h = 0 \tag{2.139}$$

2.2.2 *Gauss' Theorem*

Theorem 2.2 (Gauss' Theorem) *The integral of the outward normal component of the force* $\mathbf{F}(\mathbf{r})$ *over any closed surface containing a quantity* M *of matter is equal to* $-4\pi G M$. *In other words, the outward flux of the force through the surface is* $-4\pi G$ *times the mass enclosed (in the electrostatic case, mass is replaced by electric charge).*

Proof The integration of both sides of the Poisson's equation 2.137 over a generic subdomain N of the domain where the density is nonzero, and the application of the divergence theorem, leads to

$$\iiint\limits_N \nabla \cdot \nabla \mathrm{U}\, \mathrm{d}^3\mathbf{r} = \int\limits_{\partial \mathrm{N}} \nabla \mathrm{U} \cdot \mathbf{n}\, \mathrm{d}\sigma = -4\pi G \iiint\limits_N \rho(\mathbf{r})\, \mathrm{d}^3\mathbf{r} = -4\pi G M.$$

Exercise 2.3 Check whether the following force fields are solenoidal in $\mathbb{R}^3$, and discuss possible singular points.

(*a*) $\mathbf{F} = \dfrac{x}{y^2+z^2}\mathbf{i} + \dfrac{1}{z}\arctan\left(\dfrac{y}{z}\right)\mathbf{j} + \dfrac{1}{\sqrt{y^2+z^2}}\mathbf{k},$

(*b*) $\mathbf{F} = -\dfrac{y}{r^2}\mathbf{i} + \dfrac{x}{r^2}\mathbf{j},$

(*c*) $\mathbf{F} = \mathrm{e}^r\mathbf{r}.$

Solution To check whether the fields are solenoidal it is necessary to verify if, and where, $\nabla \cdot \mathbf{F} = 0$.

Part (*a*)

The field $\mathbf{F}(\mathbf{r})$ is singular on the plane $z = 0$ and on the entire x-axis ($y = z = 0$). With regard to solenoidality, we see that

$$\nabla \cdot \mathbf{F} = \nabla \cdot \left(\frac{x}{y^2+z^2}\mathbf{i} + \frac{1}{z}\arctan\left(\frac{y}{z}\right)\mathbf{j} + \frac{1}{\sqrt{y^2+z^2}}\mathbf{k}\right) =$$
$$= \frac{\partial}{\partial x}\left(\frac{x}{y^2+z^2}\right) + \frac{\partial}{\partial y}\left(\frac{1}{z}\arctan\left(\frac{y}{z}\right)\right) + \frac{\partial}{\partial z}\left(\frac{1}{\sqrt{y^2+z^2}}\right) =$$
$$= \frac{1}{y^2+z^2} + \frac{1}{z}\frac{z}{y^2+z^2} + \frac{-z}{(y^2+z^2)^{\frac{3}{2}}} = \frac{2\sqrt{y^2+z^2} - z}{(y^2+z^2)^{\frac{3}{2}}},$$

which vanishes only when for $z = 2\sqrt{y^2+z^2}$, and so $\mathbf{F}$ is not solenoidal.

Part (*b*)

The field is singular in the origin ($r = 0$). Given that $r = \sqrt{x^2+y^2}$, $\mathbf{F}(\mathbf{r})$ reads as

$$\mathbf{F} = -\frac{y}{x^2+y^2}\mathbf{i} + \frac{x}{x^2+y^2}\mathbf{j}$$

and

$$\nabla \cdot \mathbf{F} = \frac{\partial}{\partial x}\left(-\frac{y}{x^2+y^2}\right) + \frac{\partial}{\partial y}\left(\frac{x}{x^2+y^2}\right) =$$
$$= \frac{2xy}{\left(x^2+y^2\right)^2} - \frac{2xy}{\left(x^2+y^2\right)^2} = 0.$$

Consequently

$$\nabla \cdot \mathbf{F} = 0, \quad \forall (x, y) \in \mathbb{R}^2 \setminus \{0, 0\}.$$

Part (*c*)

The field has no singularities.

$$\nabla \cdot \mathbf{F} = \mathrm{e}^r \nabla \cdot \mathbf{r} + \mathbf{r} \cdot \nabla \mathrm{e}^r = 3\mathrm{e}^r + \mathrm{e}^r \mathbf{r} \cdot \nabla r = \mathrm{e}^r (3 + r) \neq 0, \quad \forall r,$$

and the field is not solenoidal. However, it is possible to *scale* it by a scalar function $\alpha(r)$ to make it solenoidal. Actually,

$$\nabla \cdot \alpha \mathbf{F} = \alpha \nabla \cdot \mathbf{F} + \mathbf{F} \cdot \nabla \alpha = \alpha \nabla \cdot \mathbf{F} + \frac{d\alpha}{dr} \mathbf{F} \cdot \mathbf{e}_r = 0,$$

so that

$$\frac{1}{\alpha} \frac{\mathrm{d}\alpha}{\mathrm{d}r} = -\frac{\nabla \cdot \mathbf{F}}{\mathbf{F} \cdot \mathbf{e}_r} = -\frac{\mathrm{e}^r (3 + r)}{\mathrm{e}^r r} = -\frac{3 + r}{r},$$

which implies

$$\alpha = \frac{c}{r^3 \mathrm{e}^r},$$

where $c \neq 0$ is an arbitrary constant. The field $\alpha \mathbf{F} = \dfrac{c}{r^3} \mathbf{r}$ is thus solenoidal for $r \neq 0$.

Exercise 2.4 Compute the flux, Φ, through the unitary radius sphere centred at the coordinate origin, $S_1(0)$, of the force fields:

(*a*) **F**=**r**,
(*b*) **F**=x **i**.

Solution Part (*a*)

$$\Phi = \int\limits_{\partial S_1(0)} \mathbf{F} \cdot \mathbf{n} \, \mathrm{d}\sigma = \int\limits_{\partial S_1(0)} r \, \mathrm{d}\sigma = \int\limits_0^{\pi} \int\limits_0^{2\pi} \sin \varphi \, \mathrm{d}\theta \mathrm{d}\varphi = 4\pi.$$

Part (*b*)

$$\Phi = \int\limits_{\partial S_1(0)} \mathbf{F} \cdot \mathbf{n} \, \mathrm{d}\sigma = \int\limits_{\partial S_1(0)} \frac{x^2}{r} \, \mathrm{d}\sigma = \int\limits_0^{\pi} \int\limits_0^{2\pi} \sin^2 \varphi \cos^2 \theta \sin \varphi \, \mathrm{d}\theta \, \mathrm{d}\varphi =$$

$$= \int\limits_0^{2\pi} \cos^2 \theta \, \mathrm{d}\theta \int\limits_0^{\pi} \sin^3 \varphi \, d\varphi,$$

but

$$\int\limits_0^{\pi} \sin^3 \varphi \, \mathrm{d}\varphi = -\int\limits_1^{-1} \sin^2 \varphi \, \mathrm{d}\cos \varphi = \int\limits_{-1}^{1} (1 - \cos^2 \varphi) \, \mathrm{d}\cos \varphi = -\frac{2}{3} + 2 = \frac{4}{3},$$

which gives

$$\Phi = \frac{4}{3}\int_0^{2\pi} \cos^2\theta\, \mathrm{d}\theta = \int_0^{2\pi} \left(\frac{1+\cos 2\theta}{2}\right) \mathrm{d}\theta = \frac{1}{2}\big[\theta + \sin\theta\cos\theta\big]_0^{2\pi} = \frac{4}{3}\pi.$$

Clearly, the same result can be obtained much faster via the divergence theorem

$$\Phi = \int_{S_1(0)} \nabla\cdot\mathbf{F}\,\mathrm{d}V = V = \frac{4}{3}\pi.$$

2.2.3 Gravitational Potential Energy

As we said in Sect. 2.2, the Newtonian gravitational force is a conservative field. Indeed

$$\nabla\wedge\mathbf{F}(\mathbf{r}) = G\iiint_{C} \nabla\wedge\frac{\mathbf{r}'-\mathbf{r}}{\|\mathbf{r}'-\mathbf{r}\|^3}\rho(\mathbf{r}')\,\mathrm{d}^3\mathbf{r}' = 0, \tag{2.140}$$

as can be easily verified since, for a point external to the mass domain C ($\mathbf{r}\neq\mathbf{r}'$):

$$\begin{aligned}\nabla\wedge\frac{\mathbf{r}'-\mathbf{r}}{\|\mathbf{r}'-\mathbf{r}\|^3} &= \frac{1}{\|\mathbf{r}'-\mathbf{r}\|^3}\nabla\wedge(\mathbf{r}'-\mathbf{r}) - (\mathbf{r}'-\mathbf{r})\wedge\nabla\frac{1}{\|\mathbf{r}'-\mathbf{r}\|^3} = \\ &= -(\mathbf{r}'-\mathbf{r})\wedge\left(\frac{-3}{\|\mathbf{r}'-\mathbf{r}\|^4}\frac{\mathbf{r}-\mathbf{r}'}{\|\mathbf{r}'-\mathbf{r}\|}\right) = 0.\end{aligned} \tag{2.141}$$

If, instead, the point where we calculate the curl of $\mathbf{F}$ belongs to C, we can see, in the same way followed to obtain Poisson's equation, that

$$\nabla\wedge\mathbf{F}(\mathbf{r}) = \lim_{h\to 0} G\iiint_{S_h(\mathbf{r})} \nabla\wedge\frac{\mathbf{r}'-\mathbf{r}}{\|\mathbf{r}'-\mathbf{r}\|^3}\rho(\mathbf{r}')\,\mathrm{d}^3\mathbf{r}' = 0. \tag{2.142}$$

Given its conservativity, it is possible to obtain useful information on $\mathbf{F}(\mathbf{r})$ from its potential function $U(\mathbf{r})$ (see Eq. 2.126).

We now introduce the definition of gravitational potential energy of a regular distribution of matter.

Definition 2.3 Let C be a material domain characterized by a regular density distribution $\rho(\mathbf{r})$ vanishing out of C. We define as gravitational energy, Ω, of the matter system C the work done against gravitational forces to bring all the matter elements from infinite distance to their actual configuration in C.

Let us examine two different methods to derive the expression for Ω.

The work done against the force $\mathbf{F}$ deriving from a potential $U(\mathbf{r})$ in transferring a quantity of matter δm from infinity to position $\mathbf{r}$ is

$$\delta\Omega = -\int_{\infty}^{\mathbf{r}} \mathrm{d}L = -\delta m \int_{\infty}^{\mathbf{r}} \mathbf{F}\cdot\mathrm{d}\mathbf{r} = -\delta m\left[U(\mathbf{r}) - \lim_{\mathbf{r}\to\infty} U(\mathbf{r})\right] = -\delta m\, U(\mathbf{r}), \tag{2.143}$$

given the condition $\lim_{r\to\infty} U(\mathbf{r}) = 0$. Consequently, a local infinitesimal density increment $\delta\rho(\mathbf{r})$ corresponds to an increment of potential energy

$$\delta\Omega = -\iiint_C U(\mathbf{r})\delta\rho(\mathbf{r})\,\mathrm{d}^3\mathbf{r}. \tag{2.144}$$

Due to the increment $\delta\rho$, Poisson's equation becomes

$$\nabla^2(U + \delta U) = -4\pi G(\rho + \delta\rho), \tag{2.145}$$

and its linearity implies that

$$\nabla^2\delta U = -4\pi G\delta\rho, \tag{2.146}$$

and so

$$\delta\rho(\mathbf{r}) = -\frac{1}{4\pi G}\nabla^2\delta U. \tag{2.147}$$

Substituting into Eq. 2.144 gives

$$\delta\Omega = \frac{1}{4\pi G}\iiint_C U\nabla^2\delta U\,\mathrm{d}^3\mathbf{r} \equiv \frac{1}{4\pi G}\iiint_C U\nabla\cdot\nabla\delta U\,\mathrm{d}^3\mathbf{r}. \tag{2.148}$$

If now we apply the formula 1.44, where we let $f = U$ and $\mathbf{v} = \nabla\delta U$, and applying the divergence theorem to the integral in Eq. 2.148, we get

$$\delta\Omega = \frac{1}{4\pi G}\left(\int_{\partial\mathbb{R}^3} U\nabla\delta U\cdot\mathbf{n}\,\mathrm{d}\sigma - \iiint_{\mathbb{R}^3}\nabla\delta U\cdot\nabla U\,\mathrm{d}^3\mathbf{r}\right). \tag{2.149}$$

The first integral in the above expression vanishes because, for large r

$$\begin{cases} U(r) \propto \dfrac{1}{r}, \\ |\nabla \delta U(r)| \propto \dfrac{1}{r^2}, \\ \mathrm{d}\sigma \propto r\mathrm{d}r, \end{cases} \tag{2.150}$$

so that

$$\int\limits_{\partial \mathbb{R}^3} U \nabla \delta U \cdot \mathbf{n}\, \mathrm{d}\sigma \propto \lim_{r\to\infty} \left(-\frac{1}{r} \right) = 0. \tag{2.151}$$

Consequently, Eq. 2.149 reduces to

$$\delta \Omega = -\frac{1}{4\pi G} \iiint\limits_{\mathbb{R}^3} \nabla \delta U \cdot \nabla U \, \mathrm{d}^3\mathbf{r} = -\frac{1}{4\pi G} \frac{1}{2} \delta \iiint\limits_{\mathbb{R}^3} \nabla U \cdot \nabla U \, \mathrm{d}^3\mathbf{r}. \tag{2.152}$$

Summing over all contributions $\delta \Omega$ finally yields

$$\Omega = -\frac{1}{8\pi G} \iiint\limits_{\mathbb{R}^3} |\nabla U|^2 \, \mathrm{d}^3\mathbf{r}. \tag{2.153}$$

An alternative form for Ω is obtained by using again Eq. 1.44 with the positions $f = U$ and $\mathbf{v} = \nabla U$, together with the divergence theorem, to have

$$\Omega = -\frac{1}{8\pi G} \left[\int\limits_{\partial \mathbb{R}^3} U \nabla U \cdot \mathbf{n}\, \mathrm{d}\sigma - \iiint\limits_{\mathbb{R}^3} U \nabla^2 U \, \mathrm{d}^3\mathbf{r} \right] = -\frac{1}{2} \iiint\limits_{C} U(\mathbf{r}) \rho(\mathbf{r}) \, \mathrm{d}^3\mathbf{r}, \tag{2.154}$$

the last equality being justified by that the first integral in this expression vanishes, and $\nabla^2 U = -4\pi G \rho(\mathbf{r})$.

A much faster way to get the above expression of the gravitational potential energy of a distribution of matter contained in a domain C is by using its definition as generalization to a double integral of the double sum of Eq. 2.32, that is

$$\Omega = -\frac{1}{2} G \iint\limits_{C\,C} \frac{\mathrm{d}m \, \mathrm{d}m'}{\|\mathbf{r}' - \mathbf{r}\|} = -\frac{1}{2} \int\limits_{C} \mathrm{d}m \int\limits_{C} G \frac{\mathrm{d}m'}{\|\mathbf{r}' - \mathbf{r}\|} = -\frac{1}{2} \iiint\limits_{C} U(\mathbf{r}) \rho(\mathbf{r}) \mathrm{d}^3\mathbf{r}. \tag{2.155}$$

Exercise 2.5 Obtain the expression for the gravitational potential energy of a homogeneous sphere of density ρ_0 and radius R.

Solution

$$\Omega = -\frac{1}{2}\int\limits_{S_R(0)} U(r)\,\mathrm{d}m = -\pi G\rho_0 \int\limits_0^R \left(R^2 - \frac{r^2}{3}\right)\rho_0 4\pi r^2\,\mathrm{d}r =$$

$$= -4\pi^2 G\rho_0^2\left(\frac{R^5}{3} - \frac{R^5}{15}\right) = -\frac{16\pi^2}{15}G\rho_0^2 R^5 = -\frac{3}{5}\frac{GM^2}{R},$$

where $M = (4\pi)/3)\rho_0 R^3$ is the mass of the sphere.

Note that, for a generic density distribution in spherical symmetry cut at a given radius R, the gravitational potential energy takes the form $\Omega = -A^2(GM^2)/R$, where $A^2 \simeq 1/2$ is a coefficient which depends on the "shape" of the distribution, being larger for more concentrated density profiles.

2.2.4 *Newton's Theorems*

Let us briefly discuss the classical Newton's Shell Theorems about gravitational potential and force.

Theorem 2.3 (Newton's First Theorem) *The gravitational force internal to a spherical, homogeneous, and infinitely thin material shell is null.*

Proof The theorem can be proved by noting (see Fig. 2.5) that on any point P internal to the infinitely thin spherical shell the diametrically opposite matter portions of shell, A_1 and A_2, intersected by the same solid angle in P, indicated by α, have masses Δm_1 and Δm_2 which scale with the same square power of the distance to the point as that of the (inverse) square scaling of their (inward directed and so opposite each other) gravitational force on the point, ΔF_1 and ΔF_2. That is, $\Delta m_1 = \alpha\sigma r_1^2$ and $\Delta m_2 = \alpha\sigma r_2^2$, where σ is the (constant) surface mass density, so

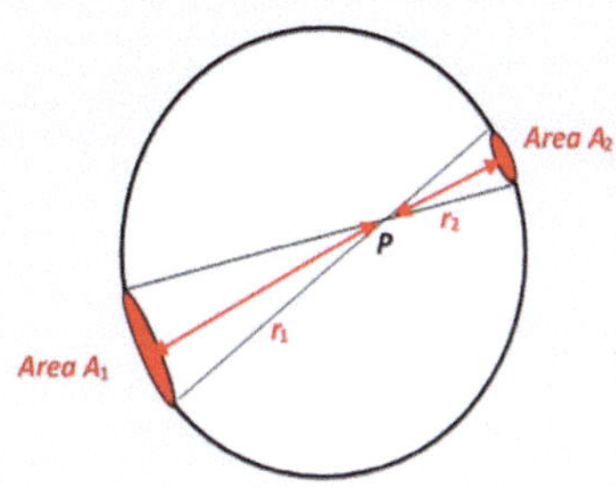

Fig. 2.5 In the internal point P the contribution of the 2 given spherical shells to the force cancels out

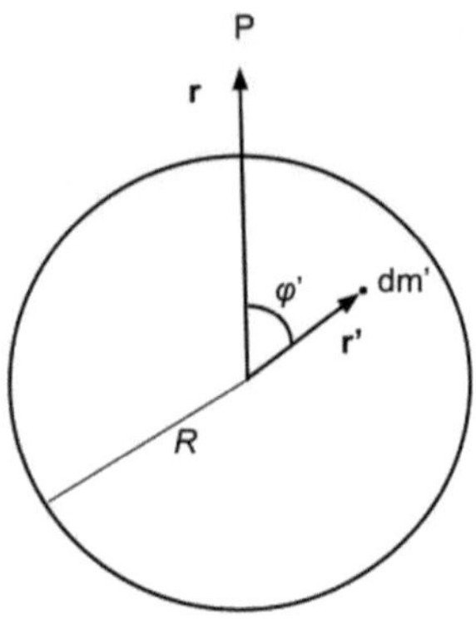

Fig. 2.6 Schematic representation of the infinitesimal element of mass dm' on the spherical surface, its radius vector $\mathbf{r}'$ ($\|\mathbf{r}'\| = R$) and the position, $\mathbf{r}$, of the external point P

that

$$\frac{\Delta F_1}{\Delta F_2} = \frac{\Delta m_1}{r_1^2}\frac{r_2^2}{\Delta m_2} = 1.$$

The proof of the Newton's first theorem results, also, as a direct corollary of Newton's second theorem:

Theorem 2.4 (Newton's Second Theorem) *The gravitational force external to a spherical, homogeneous and infinitely thin material shell is the same exerted by all of its mass as concentrated at a point at its centre.*

Proof Let P be a point of unitary mass out of the material shell of radius R and constant surface density σ, identified by its position vector $\mathbf{r}$.

Let us compute explicitly the force vector $\mathbf{F}(\mathbf{r})$. The infinitesimal mass element, dm', located in a generic point $\mathbf{r}'$ on the shell surface is given by $dm' = \sigma R \sin\varphi' d\theta' R d\varphi'$, using spherical polar coordinates with origin in the centre of the shell (see Fig. 2.6), so we have

$$\mathbf{F}(\mathbf{r}) = G\int_0^{\pi}\int_0^{2\pi} \frac{\mathbf{r}'-\mathbf{r}}{\|\mathbf{r}'-\mathbf{r}\|^3}\sigma R^2 \sin\varphi' d\theta' d\varphi' =$$

$$= 2\pi G R^2 \sigma(-\mathbf{e}_r)\int_0^{\pi} \frac{(r - R\cos\varphi')\sin\varphi'\, d\varphi'}{(R^2 + r^2 - 2rR\cos\varphi')^{3/2}},$$

where the last equality follows from noting that, by symmetry, the force points to the centre (direction $-\mathbf{e}_r$), and $(\mathbf{r}'-\mathbf{r})\cdot\mathbf{e}_r = R\cos\varphi' - r$.

Letting $D \equiv \|\mathbf{r}'-\mathbf{r}\|$, we have

$$\|\mathbf{r}'-\mathbf{r}\|^2 = R^2 + r^2 - 2rR\cos\varphi' = D^2,$$

from which it follows

$$R\cos\varphi' = \frac{R^2 + r^2 - D^2}{2r} \qquad \Longrightarrow r - R\cos\varphi' = \frac{r^2 + D^2 - R^2}{2r}.$$

The integration variable φ' can be substituted by D, taking into account that

$$\mathrm{d}D^2 = 2D\mathrm{d}D = 2rR\sin\varphi'\,\mathrm{d}\varphi' \Longrightarrow \sin\varphi'\,\mathrm{d}\varphi' = \frac{D}{rR}\,\mathrm{d}D,$$

so that, for $r > R$, the force can be written as

$$\begin{aligned}\mathbf{F}(\mathbf{r}) &= G\pi\frac{R}{r^2}\sigma(-\mathbf{e}_r)\int\limits_{r-R}^{r+R}\left(1 + \frac{r^2 - R^2}{D^2}\right)\mathrm{d}D = \\ &= G\pi\frac{R}{r^2}\sigma(-\mathbf{e}_r)\left[D - \frac{r^2 - R^2}{D}\right]_{r-R}^{r+R} = \frac{4\pi G\sigma R^2}{r^2}(-\mathbf{e}_r) = \\ &= -\frac{GM}{r^2}\mathbf{e}_r,\end{aligned} \tag{2.156}$$

where $M = 4\pi\sigma R^2$ is the shell mass.

Although, historically, Newton gave the demonstration of what is called Newton's first theorem by geometrical arguments before introducing his second theorem, its thesis is easily derived as a corollary of his second theorem, as we now see.

Corollary 2.1 (Demonstration of Newton's First Theorem)

Proof For an internal point ($r < R$), the lower limit of integration in Eq. 2.156 changes: D runs from $R - r$ to $r + R$. Substituting these limits into Eq. 2.156 gives

$$\mathbf{F}(\mathbf{r}) = G\pi\frac{R}{r^2}\sigma(-\mathbf{e}_r)\int\limits_{R-r}^{r+R}\left(1 + \frac{r^2 - R^2}{D^2}\right)\mathrm{d}D = 0.$$

Another way to derive both of the Newton's shell theorems is by means of a calculation of the potential, as follows.

The gravitational potential, $U_s(\mathbf{r})$, of a spherically symmetric material shell is given by the double integral

$$U_s(\mathbf{r}) = G\int\limits_0^{\pi}\int\limits_0^{2\pi}\frac{\sigma R\sin\varphi\,\mathrm{d}\theta\,R\,\mathrm{d}\varphi}{\sqrt{R^2 + r^2 - 2rR\cos\varphi}} =$$

$$= GR^2 \frac{2\pi\sigma}{2rR} \int_{-2rR}^{2rR} \frac{\mathrm{d}y}{\sqrt{R^2 + r^2 - y}} = -G\frac{R}{r} 2\pi\sigma \left[\sqrt{R^2 + r^2 - y}\right]_{-2rR}^{2rR} =$$

$$= -2\pi G \frac{R}{r}\sigma\Big[|R - r| - (R + r)\Big] = \begin{cases} 4\pi G \dfrac{R^2}{r}\sigma \text{ if } r \geq R \\ 4\pi G R\sigma \quad \text{if } r < R, \end{cases}$$

so that the force is

$$\mathbf{F}(\mathbf{r}) = \nabla U_s = \begin{cases} -4\pi G \dfrac{R^2}{r^2}\sigma\, \mathbf{e}_r, \text{ if } r \geq R, \\ 0, \qquad\qquad \text{if } r < R, \end{cases} \tag{2.157}$$

which is equivalent to the results of first and second Newton's theorem.

2.3 Further Readings

The classical, Newtonian, theory of gravitation is one of the most studied, and applied, fields in theoretical and experimental Physics. Consequently, there is an enormous quantity of textbooks which can help the curious reader to broaden and deepen the notions presented in this chapter. As mere examples, I suggest [4, 6, 8]. Some deep theoretical insight into gravitation is found in the 2-volume book [3], with extension to General Relativity in the fundamental texts [7, 10]. A careful astrophysical perspective comes from reading [2, 9, 11]. Deepening on the study of N-body systems is found in [5]. The role of numerical simulations in the context of N-body dynamics is addressed in [1].

References

1. Aarseth, S.J.: Gravitational N-Body Simulations, Tools and Algorithms. Cambridge Monographs on Mathematical Physics, Cambridge University Press, Cambridge (2001)
2. Binney, J., Tremaine, S.: Galactic Dynamics, 2nd edn. Princeton University Press, Princeton (2011)
3. Fridman, A.M., Polyachenko, V.L.: Physics of Gravitating Systems I and II. Springer, Berlin (2012)
4. Goldstein, H., Safko, J.L., Poole, G.P., Spiegel, M.R.: Classical Mechanics, 3rd edn. Pearson Ed. Limited, Edinburgh (2002)

5. Heggie, D.C., Hut, P.: The Gravitational Million Body Problem: A Multidisciplinary Approach to Star Cluster Dynamics. Cambridge University Press, Cambridge (2003)
6. Landau, L.D., Lifshits, E.M.: Mechanics, Course of Theoretical Physics, vol. 1, 3rd edn. Elsevier Butterworth-Heinemann, Oxford (1976)
7. Landau, L.D., Lifshits, E.M.: The Classical Theory of Fields, Course of Theoretical Physics, vol. 2, 4th edn. Elsevier Butterworth-Heinemann, Oxford (1987)
8. Mc Call, M.W.: Classical Mechanics: From Newton to Einstein: A Modern Introduction, 2nd edn. J. Wiley & Sons Ltd., Chichester (2011)
9. Merritt, D.: Dynamics and Evolution of Galactic Nuclei. Princeton University Press, Princeton (2013)
10. Misner, C.W., Thorne, K.S., Wheeler, J.A.: Gravitation. W. H. Freeman and Company, New York (1973)
11. Saslaw, W.C.: Gravitational Physics of Stellar and Galactic Systems. Cambridge University Press, Cambridge (1985)
12. Wang, Q.D., The global solution of the N-body problem. Celestial Mech. Dyn. Astron. **50**(1), 73–88 (1991)

Chapter 3
Central Force Fields

This chapter studies the characteristics of *central* force fields, i.e. those force vector fields which point towards a fixed point. This study is of particular relevance for Newtonian gravity, because a spherical distribution of matter determines a central force field in space.

After introducing the topic, by means of Poisson's equation and the fundamentals of potential theory we obtain the expressions of the potential (and force) for simple spherical matter distributions: the point-like case, the homogeneous sphere, the Plummer sphere, the isochrone sphere.

We address the topic of the analysis of motion in such potentials both *quantitatively*, deriving trajectories and laws of motion when possible, and *qualitatively*, delimiting the allowed phase space through the proper use of constants and integrals of motion. The possible simple and multiple periodicities of oscillations in such potentials are also investigated.

3.1 The Potential and Force Generated by a Spherical Matter Distribution

The Newtonian potential generated by matter distributed over a domain C according to a density law $\rho(\mathbf{r})$ is, by definition

$$U(\mathbf{r}) = G \int\limits_C \frac{\mathrm{d}m'}{\|\mathbf{r}' - \mathbf{r}\|} = G \iiint\limits_C \frac{\rho(\mathbf{r}')}{\|\mathbf{r}' - \mathbf{r}\|}\, \mathrm{d}^3\mathbf{r}', \tag{3.1}$$

so that if C is limited in space, then $\lim\limits_{r\to\infty} U(\mathbf{r}) = 0$.

The Euclidean distance between the two points $\mathbf{r}$ and $\mathbf{r}'$ can be written as

$$\|\mathbf{r}' - \mathbf{r}\| = \sqrt{r'^2 + r^2 - 2r'r\cos\widehat{\mathbf{r}\mathbf{r}'}}, \tag{3.2}$$

R. A. Capuzzo Dolcetta, *Classical Newtonian Gravity and Potential Theory*,
UNITEXT for Physics, https://doi.org/10.1007/978-3-032-13036-5_3

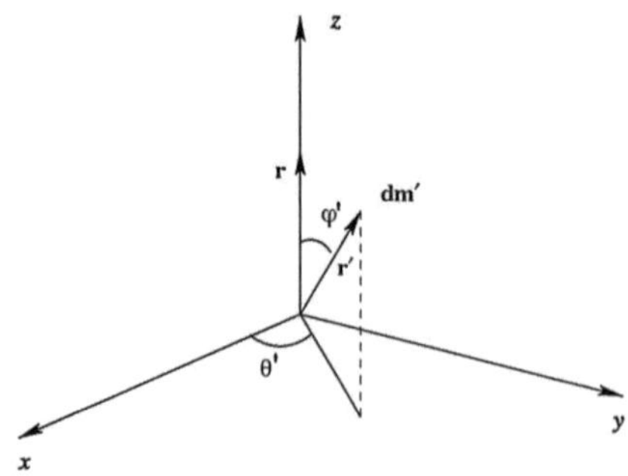

Fig. 3.1 A coordinate frame obtained after a rotation that aligns the z-axis with the $\mathbf{r}$ direction

where $\widehat{\mathbf{rr}'}$ is the angle between the vectors $\mathbf{r}$ and $\mathbf{r}'$; in spherical polar coordinates (r, θ, φ), the cosine theorem gives

$$\cos \widehat{\mathbf{rr}'} = \cos\varphi \cos\varphi' + \sin\varphi \sin\varphi' \cos(\theta - \theta'). \tag{3.3}$$

Substitution by means of (3.2) and (3.3) into Eq. 3.1, and taking into account that, in spherical polar coordinates, $\rho(\mathbf{r}')\mathrm{d}^3\mathbf{r}' = \rho(r', \theta', \varphi')r'^2 \sin\varphi' \mathrm{d}\theta' \mathrm{d}\varphi' \mathrm{d}r'$, makes evident that the integral 3.1 for the potential is very complicated to calculate.

Anyway, in the case of spherical symmetry for the density, i.e. $\rho(\mathbf{r}') = \rho(r')$, the problem is clearly invariant under rotations around the centre of symmetry, and so we can choose the polar reference frame setting the z-axis from which to evaluate the φ angles along the $\mathbf{r}$ direction, so that in this new spherical polar coordinate frame the vector $\mathbf{r}$ (Fig. 3.1) has $\varphi = 0$, yielding the simplification

$$\|\mathbf{r}' - \mathbf{r}\| = \sqrt{r'^2 + r^2 - 2r'r \cos\varphi'}. \tag{3.4}$$

In the case of a spherical domain of finite radial extension R, we have

$$\begin{aligned} U(r) &= G \int_0^R \int_0^\pi \int_0^{2\pi} \rho(r') \frac{r'^2 \sin\varphi' \mathrm{d}\theta' \mathrm{d}\varphi' \mathrm{d}r'}{\sqrt{r'^2 + r^2 - 2r'r\cos\varphi'}} = \\ &= 2\pi G \int_0^R \rho(r') r'^2 \mathrm{d}r' \int_0^\pi \frac{\sin\varphi'}{\sqrt{r'^2 + r^2 - 2r'r\cos\varphi'}} \mathrm{d}\varphi'. \end{aligned} \tag{3.5}$$

Letting $t(\varphi') = r'^2 + r^2 - 2r'r\cos\varphi'$, the above integral expression becomes

$$
\begin{aligned}
U(r) &= 2\pi G \int_0^R \rho(r')r'^2 \mathrm{d}r' \int_{t(0)}^{t(\pi)} \frac{1}{2r'r}\frac{\mathrm{d}t}{\sqrt{t}} = \\
&= 2\pi G \int_0^R \rho(r')r'^2 \frac{1}{rr'}(r' + r - |r' - r|)\mathrm{d}r'.
\end{aligned}
\tag{3.6}
$$

Now, we distinguish between the potential (i) external to the distribution ($r \geq R$), and (ii) internal ($0 \leq r \leq R$).

Actually

(i) when $r \geq R$ (and $r' \leq R \leq r$), $|r' - r| = r - r'$, so that

$$
U(r) = \frac{4\pi G}{r} \int_0^R \rho(r')r'^2 \mathrm{d}r' = G\frac{M(R)}{r} \equiv U_{int}(r);
\tag{3.7}
$$

(ii) when $0 \leq r \leq R$, it is $|r'-r| = r-r'$ for $0 \leq r' \leq r$, while it is $|r'-r| = r'-r$, for $0 \leq r \leq r'$, leading to

$$
\begin{aligned}
U(r) &= \frac{4\pi G}{r}\left[\int_0^r r'^2 \rho(r')\,\mathrm{d}r' + r\int_r^R r'\rho(r')\,\mathrm{d}r'\right] = \\
&= \frac{GM(r)}{r} + 4\pi G \int_r^R r'\rho(r')\,\mathrm{d}r' \equiv U_{int}(r) + U_{ext}(r).
\end{aligned}
\tag{3.8}
$$

Note that while the potential on a generic point P at distance r from the origin is contributed by both the matter within r ($U_{int}(r)$) and out of r ($U_{ext}(r)$), the force has only the internal contribution, as already stated by Newton's theorems. Indeed, for any choice of integrable spherically symmetric density law, the force generated by the spherical potential is:

$$
\begin{aligned}
\mathbf{F} = \nabla \mathrm{U} = \mathrm{U}'(r)\mathbf{e}_r &= \left\{ -\frac{4\pi G}{r^2}\left[\int_0^r r'^2 \rho(r')\,\mathrm{d}r' + r\int_r^R r'\rho(r')\,\mathrm{d}r'\right] + \right. \\
&\left. + \frac{4\pi G}{r}\left[\cancel{r^2\rho(r)} + \int_r^R r'^2 \rho(r')\,\mathrm{d}r' - \cancel{r^2\rho(r)}\right]\right\}\mathbf{e}_r = -\frac{GM(r)}{r^2}\mathbf{e}_r.
\end{aligned}
\tag{3.9}
$$

Exercise 3.1 Calculate the average distance between a generic pair of particles in a homogeneous sphere of radius R.

Solution Consider two randomly chosen points $\mathbf{r}$ and $\mathbf{r}'$ within the sphere, $r \leq R$ and $r' \leq R$, in a homogeneous sphere containing N objects.

The probability density distribution of the distance, D, between the two generic points in a uniform sphere is

$$P(D) = \frac{3N}{16R^6} D^2 (4R^2 - D^2)^2,$$

consequently, the average of D is

$$\langle D \rangle = \frac{1}{N} \int_0^{2R} D P(D) \mathrm{d}D = \frac{3}{16R^6} \int_0^{2R} D \frac{D^2 (4R^2 - D^2)^2}{R^6} \mathrm{d}D,$$

independent of N. By developing the square of the binomial the result is

$$\langle D \rangle = \frac{36}{35} R \simeq 1.02857 R,$$

slightly larger than the sphere radius (as expected). Recalling the average closest neighbour distance obtained in the Exercise 2.1, we have

$$\frac{\langle D_{cn} \rangle}{\langle D \rangle} \simeq 0.86818 \sqrt[3]{N}.$$

3.1.1 *Calculating Spherical Potentials via Poisson's Equation*

The potential generated by a spherical matter distribution can be easily obtained by solving Poisson's equation in spherical polar coordinates.

Actually, if $\rho(\mathbf{r})$ is spherically symmetric, also $U(\mathbf{r})$ is spherically symmetric, so that the Poisson's equation (see Eq. 1.65) reduces to

$$\frac{1}{r^2} \frac{\mathrm{d}}{\mathrm{d}r} \left(r^2 \frac{\mathrm{d}U}{\mathrm{d}r} \right) = -4\pi G \rho(r). \tag{3.10}$$

A first integration over r of Eq. 3.10 gives

$$\int_0^r \mathrm{d} \left(r^2 \frac{\mathrm{d}U}{\mathrm{d}r} \right) = -G \int_0^r \rho(r) 4\pi r^2 \, \mathrm{d}r = -GM(r), \tag{3.11}$$

leading to

$$r^2 \frac{\mathrm{d}U}{\mathrm{d}r} - \lim_{r \to 0} r^2 \frac{\mathrm{d}U}{\mathrm{d}r} = -GM(r). \tag{3.12}$$

Now, it is $\lim_{r\to 0} r^2(\mathrm{d}U/\mathrm{d}r) = 0$, because any regular matter distribution is *broader* than that of the point-like case (which is a Dirac's *delta* function) which produces the $U \propto 1/r$ potential, thus Eq. 3.12 simply gives the first and second Newton's theorem theses, resumed as

$$\frac{\mathrm{d}U}{\mathrm{d}r} = -G\frac{M(r)}{r^2}. \tag{3.13}$$

An integration by parts (with $u = M(r)$ and $\mathrm{d}v = \mathrm{d}r/r^2$) of the equation above leads to

$$\begin{aligned}\int_r^\infty \mathrm{d}U &= -G\int_r^\infty \frac{M(r)}{r^2}\mathrm{d}r = G\left[\frac{M(r)}{r}\right]_r^\infty - G\int_r^\infty \frac{\mathrm{d}M(r)}{r} = \\ &= G\left(\lim_{r\to\infty}\frac{M(r)}{r} - \frac{M(r)}{r}\right) - 4\pi G\int_r^\infty \rho(r)r\mathrm{d}r,\end{aligned} \tag{3.14}$$

which reduces to

$$U(r) = G\frac{M(r)}{r} + 4\pi G\int_r^\infty \rho(r)r\mathrm{d}r + U_\infty, \tag{3.15}$$

letting $U_\infty = \lim_{r\to\infty} U(r)$ and assuming $\lim_{r\to\infty} M(r)/r = 0$. Note that the condition $\lim_{r\to\infty} M(r)/r = 0$ means that the asymptotic (large r) behaviour of $\rho(r)$ is $\rho \propto 1/r^{2+\beta}$, with $\beta > 0$. This asymptotic behaviour for $\rho(r)$ guarantees, also, convergence of the integral in the rhs of Eq. 3.15. Finally, note that for a finite mass distribution (i.e. $\rho(r) = 0$ for $r > R$), the integral in Eq. 3.15 extends only to R, and the equation reduces to Eq. 3.8 provided that $U_\infty = 0$.

Exercise 3.2 Compute the flux, Φ, through a spherical surface of radius R of the gravitational force, $\mathbf{F}(r)$, exerted by a spherical matter distribution, $\rho(r)$.

Solution The flux is given by

$$\Phi = \int_{S_R(0)} \mathbf{F}(r)\cdot\mathbf{n}\,\mathrm{d}\sigma = -\int_0^\pi\int_0^{2\pi} \frac{GM(R)}{R^2}R^2\sin\varphi\,\mathrm{d}\theta\mathrm{d}\varphi = -4\pi GM(R),$$

as expected by the Gauss' law in Eq. 1.109.

3.2 Motion in a Spherical Potential

As we saw in Sect. 1.4 of Chap. 1, in the case of a central force field per unit mass, $\mathbf{F}(\mathbf{r}) = f(\mathbf{r})\,\mathbf{e}_r$, the orbital angular momentum per unit mass, $\mathbf{L}$, is conserved because

$$\dot{\mathbf{L}} = \mathbf{r} \wedge f(\mathbf{r})\,\mathbf{e}_r = 0, \tag{3.16}$$

so that the motion is planar.

In such force field, the particle acceleration, in polar coordinates (r, θ) defined on the motion plane, takes the form

$$\mathbf{a} = (\ddot{r} - r\dot{\theta}^2)\,\mathbf{e}_r + (r\ddot{\theta} + 2\dot{r}\dot{\theta})\,\mathbf{e}_\theta, \tag{3.17}$$

and the equation of motion in scalar form reduces to

$$\begin{cases} \ddot{r} - r\dot{\theta}^2 = f(r), \\ r\ddot{\theta} + 2\dot{r}\dot{\theta} = 0, \end{cases} \tag{3.18}$$

where we have also assumed spherical symmetry for $f(\mathbf{r})$, to guarantee conservativity, as shown in Exercise 1.27.

The second equation in the system 3.18 is equivalent to

$$\frac{d}{dt}\left(r^2\dot{\theta}\right) = 0, \tag{3.19}$$

that gives $L_z = r^2\dot{\theta} = const.$, i.e. the conservation of the only nonzero component of the orbital angular momentum (conservation already guaranteed because of Eq. 3.16).

Note that L_z can be either positive or negative. The $L_z > 0$ case corresponds to *prograde* (counterclockwise) motion ($\dot{\theta} > 0$), while $L_z < 0$ to *retrograde* (clockwise) motion ($\dot{\theta} < 0$). It is important to note that the characteristic of a motion to be prograde or retrograde is given by the initial condition for $\dot{\theta}$, and, due to angular momentum conservation (no flip of $\mathbf{L}$ can occur in a central force field), this characteristic cannot be changed unless an external torque is applied.

Without any loss of generality, hereafter we take as positive direction of the z-axis orthogonal to the plane of motion the same of the angular momentum $\mathbf{L}$, so to identify L_z with $L = \|\mathbf{L}\|$.

The angular momentum conservation allows the elimination of $\dot{\theta}$ in the first equation in 3.18, so to have the motion equation for the radial coordinate

$$\ddot{r} - \frac{L^2}{r^3} = f(r). \tag{3.20}$$

Since the force is conservative, $f(r)$ is equal to $\frac{\mathrm{d}U}{\mathrm{d}r}$, and so

$$\ddot{r} = \frac{L^2}{r^3} + \frac{\mathrm{d}U}{\mathrm{d}r} \equiv \frac{\mathrm{d}U_{eff}}{\mathrm{d}r}, \tag{3.21}$$

where the *effective* potential, $U_{eff}(r; L^2)$, is defined as

$$U_{eff}(r; L^2) = \int \frac{L^2}{r^3}\,\mathrm{d}r + U = -\frac{1}{2}\frac{L^2}{r^2} + U + c, \tag{3.22}$$

with c an integration constant which is zero under the two conditions $\lim_{r\to\infty} U(r) = 0$ and $\lim_{r\to\infty} U_{eff}(r; L) = 0$. Note that the term $-L^2/(2r^2)$ in the above equation is the *centrifugal* potential.

3.2.1 Circular Trajectories

In the above scheme, provided that the force is attractive, circular trajectories exist of any radius r, upon the proper choice of the angular momentum L. Indeed, Eq. 3.21, under the assumption $r = const.$ for every t—that implies $\dot{r} = \ddot{r} = 0$—gives for the angular momentum of the unit mass particle on the circular orbit of radius r the value

$$L_c^2 = -r^3 \frac{\mathrm{d}U}{\mathrm{d}r}. \tag{3.23}$$

Note that the existence of circular trajectories of motion stands on the attractive nature of the force, $\mathrm{d}U/\mathrm{d}r < 0$, otherwise it would be $L_c^2 < 0$.

For a potential satisfying the Poisson's equation, the equation above is equivalent to

$$L_c^2 = r\,GM(r), \tag{3.24}$$

which, in virtue of the continuity of $M(r)$ and given that $M(r)$ is non-decreasing with r such that $\lim_{r\to 0} M(r)$ is finite, has always a unique solution because $\lim_{r\to\infty} r\,GM(r) = +\infty$.

Given that $L = r^2\dot{\theta} = r v_\theta$, the circular speed, $v_c(r)$, is given by

$$v_c(r) \equiv \frac{L_c}{r} = \sqrt{-r\frac{\mathrm{d}U}{\mathrm{d}r}}. \tag{3.25}$$

In principle, circular orbits in spherical potentials may be either *stable* (if the orbital radius corresponds to a minimum of V_{eff}) or *unstable* (if it corresponds to a maximum of V_{eff}); however, if the potential has a gravitational origin (i.e. it satisfies the Poisson's equation) the unstable case is never realized (see Note 13 in Sect. 3.6).

The simplest case, that of motion in a point mass potential, gives $v_c(r) = \sqrt{(GM)/r}$, which is the *Keplerian* circular velocity law.

3.3 Potential Generated by a Homogeneous Sphere

As we saw in Sect. 3.1, the expression of the gravitational potential of a generic, spherically symmetric, distribution of matter is (Eq. 3.8)

$$U(r) = G\frac{M(r)}{r} + 4\pi G \int_r^R r\rho(r)\,\mathrm{d}r, \tag{3.26}$$

which, assuming $\rho(r) = \rho_0 = const. > 0$ for $r \leq R$, and $\rho(r) = 0$ for $r > R$, gives

$$U(r) = \begin{cases} \frac{4}{3}\pi G\rho_0 r^2 + 4\pi G\rho_0 \left[\frac{r'^2}{2}\right]_r^R = 2\pi G\rho_0 \left(R^2 - \frac{1}{3}r^2\right), & \text{if } r \leq R \\ \frac{4}{3}\pi G\rho_0 \frac{R^3}{r}, & \text{if } r \geq R, \end{cases} \tag{3.27}$$

that is

$$U(r) = \begin{cases} \text{harmonic}, & \text{for } r \leq R, \\ \text{Keplerian}, & \text{for } r \geq R. \end{cases} \tag{3.28}$$

3.3.1 *Trajectories in a Homogeneous Sphere*

As we saw above, the force inside a homogeneous sphere is harmonic, $\mathbf{F} = -kr\,\mathbf{e}_r$, where the *elastic* constant is $k = \frac{4}{3}G\pi\rho_0 \equiv \omega^2 > 0$. The second-order vector differential equation of motion is $\ddot{\mathbf{r}} = -\omega^2 r\,\mathbf{e}_r$, which, once projected onto the xy-plane of motion, gives

$$\begin{cases} \ddot{x} = -\omega^2 x \\ \ddot{y} = -\omega^2 y, \end{cases} \tag{3.29}$$

Fig. 3.2 The particle starts moving with $x_0 > 0$, $\dot{x}_0 = 0$, $y_0 = 0$, $\dot{y}_0 > 0$

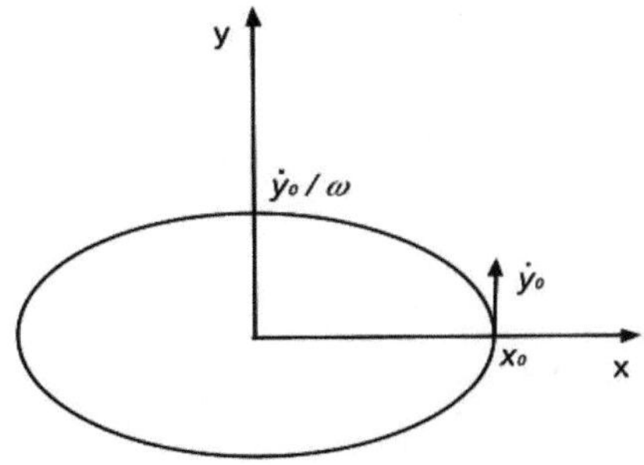

whose general solutions are $x(t) = a_x \cos\omega t + b_x \sin\omega t$ and $y(t) = a_y \cos\omega t + b_y \sin\omega t$, with $a_x,\ b_x,\ a_y,\ b_y$, constants to be determined by the initial conditions on $\mathbf{r}$ and $\dot{\mathbf{r}}$, namely $x(0) = x_0,\ \dot{x}(0) = \dot{x}_0,\ y(0) = y_0,\ \dot{y}(0) = \dot{y}_0$.

Upon the given initial conditions, the solutions specify as

$$x(t) = x_0 \cos\omega t + \frac{\dot{x}_0}{\omega}\sin\omega t, \qquad y(t) = y_0 \cos\omega t + \frac{\dot{y}_0}{\omega}\sin\omega t, \tag{3.30}$$

which are the *parametric* expressions of the trajectory.

The choice of the particular i.c. $x_0 > 0,\ \dot{x}_0 = 0,\ y_0 = 0,\ \dot{y}_0 > 0$ (see Fig. 3.2), which correspond to an initial velocity vector orthogonal to the initial position vector, assumed on its turn on the x-axis, leads, after squaring and summing member to member in Eq. 3.30, to the canonical Cartesian equation of an ellipse

$$\frac{x^2}{x_0^2} + \frac{y^2}{(\dot{y}_0/\omega)^2} = 1, \tag{3.31}$$

where $a^2 \equiv x_0^2$ and $b^2 \equiv (\dot{y}_0/\omega)^2$ are the (squared) semi axes along x and y, respectively.

Actually, we can see that the trajectories are *conics* for any choice of i.c.. Indeed, solving Eqs. 3.30 for $\cos\omega t$ and $\sin\omega t$, and using the Pythagorean identity, it results

$$\cos^2\omega t + \sin^2\omega t = 1 = \frac{(\frac{\dot{y}_0}{\omega}x - \frac{\dot{x}_0}{\omega}y)^2}{(\frac{\dot{y}_0}{\omega}x_0 - \frac{\dot{x}_0}{\omega}y_0)^2} + \frac{(x_0 y - y_0 x)^2}{(\frac{\dot{y}_0}{\omega}x_0 - \frac{\dot{x}_0}{\omega}y_0)^2}, \tag{3.32}$$

and, after some manipulations,

$$\left(\frac{\dot{y}_0^2}{\omega^2} + y_0^2\right)x^2 - 2\left(\frac{\dot{y}_0}{\omega^2}\dot{x}_0 + x_0 y_0\right)xy + \left(\frac{\dot{x}_0^2}{\omega^2} + x_0^2\right)y^2 = \frac{1}{\omega^2}\left(x_0\dot{y}_0 - y_0\dot{x}_0\right)^2. \tag{3.33}$$

The above expression is that of a conic section, whose general equation is

$$Ax^2 + Bxy + Cy^2 = D, \tag{3.34}$$

with, in our case,

$$\begin{cases} A = \dfrac{\dot{y}_0^2}{\omega^2} + y_0^2 \geq 0, \quad B = -2\left(\dfrac{\dot{y}_0}{\omega^2}\dot{x}_0 + x_0 y_0\right), \\ C = \dfrac{\dot{x}_0^2}{\omega^2} + x_0^2 \geq 0, \quad D = \dfrac{1}{\omega^2}\left(x_0\dot{y}_0 - y_0\dot{x}_0\right)^2 \geq 0. \end{cases} \tag{3.35}$$

If $A = C$ and $B = 0$ the curve is a circumference.

In general, defined the *discriminant* of the conic as $\Delta \equiv B^2 - 4AC$, it is known, from analytic geometry, that

$$\begin{cases} \Delta < 0 & \text{for ellipse,} \\ \Delta = 0 & \text{for parabola,} \\ \Delta > 0 & \text{for hyperbola.} \end{cases} \tag{3.36}$$

In our case

$$\Delta = -4\left(\frac{\dot{y}_0}{\omega}x_0 - \frac{\dot{x}_0}{\omega}y_0\right)^2 = -4D \leq 0, \tag{3.37}$$

so that trajectories are either ellipses ($D > 0$) or parabolas ($D = 0$), these latter being degenerate. Actually, $D = 0 \Longleftrightarrow (x_0\dot{y}_0 - y_0\dot{x}_0)^2 = \|\mathbf{r}_0 \wedge \mathbf{v}_0\|^2 = L_0^2 = 0$, meaning that $\mathbf{r}_0 \parallel \mathbf{v}_0$. Due to that $\mathbf{L}$ is constant, it results $\mathbf{r}(t) \parallel \mathbf{v}(t)$ for every t, that means the parabola degenerates in two coinciding straight lines (purely radial motion).

Formally

$$\left(\frac{\dot{y}_0^2}{\omega^2} + y_0^2\right)x^2 + \left(\frac{\dot{x}_0^2}{\omega^2} + x_0^2\right)y^2 - 2\left(\frac{\dot{y}_0}{\omega^2}\dot{x}_0 + x_0 y_0\right)xy = 0 \tag{3.38}$$

equivalent to

$$\left(\frac{\dot{y}_0}{\omega}x - \frac{\dot{x}_0}{\omega}y\right)^2 + (y_0 x - x_0 y)^2 = 0, \tag{3.39}$$

which reduces to

$$\begin{cases} \dfrac{\dot{y}_0}{\omega}x - \dfrac{\dot{x}_0}{\omega}y = 0, \\ y_0 x - x_0 y = 0. \end{cases} \tag{3.40}$$

The two straight lines above are coincident because $\dot{y}_0/\dot{x}_0 = y_0/x_0$.

3.3.2 Radial Motion in a Homogeneous Sphere

As said above, the case of i.c. such that $L_0 = 0$ leads, as always in a central force field, to a purely radial motion.

Remembering the variable separation introduced in Sect. 3.2, leading to Eq. 3.21 where we let $L = 0$, it is

$$\ddot{r} = \frac{\mathrm{d}U}{\mathrm{d}r} = -\frac{4}{3}\pi G\rho_0 r = -\omega^2 r, \tag{3.41}$$

whose solution is a *harmonic* radial oscillation across the origin for negative mechanical energy, E, and an unlimited straight line motion for $E \geq 0$ (see Sect. 3.5.2). In the $E < 0$ case, the *radial period* of oscillation, T_r, defined as the time such that $r(t + T_r) = r(t), \forall t$, that is

$$T_r = \frac{1}{2}\frac{2\pi}{\omega} = \frac{\pi}{\sqrt{\frac{4}{3}\pi G\rho_0}} = \frac{1}{2}\sqrt{\frac{3\pi}{G\rho_0}}. \tag{3.42}$$

Notably, T_r is *independent* of both $r(0) \leq R$ and $\dot{r}(0)$.

Exercise 3.3 Calculate the time needed for a person who falls into an imaginary tunnel to cross diametrically the Earth to reach the antipode. Compare this time with that required by a normal line aircraft to reach the same point.

Solution The sought time is simply given by Eq. 3.42 assuming the Earth as a homogeneous sphere of density $\rho_0 = 5.5$ g cm^{-3} and radius $R = 6370$ km. The air resistance in the tunnel is neglected as any other energy dissipation, as is the Earth rotation. The result is that the time to fall across the Earth and reach the antipode is $\sim$42.2 min, i.e. $\sim$0.70 hr, while the time for the aircraft is $\sim$22.23 hr. This means that the *free fall* in the tunnel is $\sim$31.6 times faster than the air travel.

3.4 Some Relevant Spherical Models

As we will see in Chap. 4, in the class of spherical potentials, all admissible equilibrium mass density distributions should be less concentrated than the, extreme, point mass case (Dirac's delta function for the density, $\rho(\mathbf{r}) = \delta(\mathbf{r})$) while typically keeping a radially decreasing profile $\mathrm{d}\rho/\mathrm{d}r < 0$. So the density distributions, and the corresponding potentials, should lie between the two extreme cases of Keplerian (point mass) and harmonic (homogeneous sphere) cases, which are the only spherical potentials giving bound trajectories that are closed (fixed, non-precessing). This statement is, basically, the thesis of *Bertrand's theorem* [2], whose demonstration is rather complicated and beyond the scope of this book; anyway, we will present in Sect. 5.17 the exercise 5.13 whose solution constitutes a partial demonstration of the theorem.

A direct consequence of the Bertrand's theorem is that in any realistic model of stationary, stable, spherical self-gravitating system (but the 2 cases above mentioned) the bound orbits are precessing. It may be useful, so, to study some spherical distributions that are intermediate between the two extreme cases of point-mass and uniform density. Two relevant examples are the so-called *Plummer* and *isochrone* potentials.

3.4.1 The Plummer Sphere

The Plummer (or Schuster, see [3]) sphere is an unbound spherical distribution of matter whose density law is

$$\rho(r) = \frac{\rho_0}{\left[1+\left(\frac{r}{b}\right)^2\right]^{5/2}}, \tag{3.43}$$

where ρ_0 is the central density, and $b > 0$ is a length scale. Note that $\rho(b) = \rho_0/2^{5/2}$.

It is worth noting that the above density law is an exact solution of the Lane-Emden equation. The Lane-Emden equation (see also the following Sect. 4.2.1 in Chap. 4) is the radially–symmetric equilibrium equation for a self-gravitating, polytropic, matter distribution. A *polytrope* is a matter distribution where density and pressure are related in a power law form, of the type $p = k\rho^{1+1/n}$, where $k > 0$ and $n > 0$ are constants, the latter called polytropic index. Exact solutions of the Lane-Emden equation exist only for $n = 0$, $n = 1$ and $n = 5$, this latter corresponding to the Plummer density law above.

Another relevant aspect of the Plummer density distribution is that, upon projection, it constitutes a good fit for the light distribution of globular star clusters[1] up to distances not too far from the centre.

The mass contained in a sphere of radius r centred in the symmetry centre is (letting $x = r/b$)

$$M(r) = 4\pi\rho_0 \int\limits_0^r \frac{r^2}{\left[1+\left(\frac{r}{b}\right)^2\right]^{5/2}}\,\mathrm{d}r = 4\pi\rho_0 b^3 \int\limits_0^{r/b} \frac{x^2}{\left(1+x^2\right)^{5/2}}\,\mathrm{d}x. \tag{3.44}$$

[1] Globular clusters are star cluster satellites of our Milky Way, composed by 10^5–10^6 stars distributed in a nearly spherical volume of diametral size $\sim$100 pc.

Adopting $u = x$ and $\mathrm{d}v = \dfrac{x}{(1+x^2)^{\frac{5}{2}}}\mathrm{d}x$ in the usual formula of integration by parts, we have

$$M(r) = 4\pi\rho_0 b^3 \left\{ \left[-\frac{1}{3}\frac{x}{(1+x^2)^{3/2}} \right]_0^{r/b} + \frac{1}{3}\int_0^{r/b} \frac{\mathrm{d}x}{(1+x^2)^{3/2}} \right\}. \tag{3.45}$$

By addition and subtraction of x^2 in the numerator of the integral in the rhs of Eq. 3.45, this integral becomes

$$\int_0^{r/b} \frac{\mathrm{d}x}{(1+x^2)^{3/2}} = \int_0^{r/b} \frac{\mathrm{d}x}{\sqrt{1+x^2}} - \int_0^{r/b} \frac{x^2}{(1+x^2)^{3/2}}\,\mathrm{d}x. \tag{3.46}$$

By integration by parts of the second integral in the rhs of Eq. 3.46 we have

$$\begin{aligned} \int_0^{r/b} \frac{\mathrm{d}x}{(1+x^2)^{3/2}} &= \int_0^{r/b} \frac{\mathrm{d}x}{\sqrt{1+x^2}} - \left[\frac{1}{2}\frac{(-2)\,x}{\sqrt{1+x^2}} \right]_0^{r/b} + \int_0^{r/b} \frac{\mathrm{d}x}{\sqrt{1+x^2}} = \\ &= \frac{r/b}{\sqrt{1+\left(\frac{r}{b}\right)^2}}. \end{aligned} \tag{3.47}$$

Finally

$$\begin{aligned} M(r) &= 4\pi\rho_0 b^3 \left\{ -\frac{1}{3}\frac{r/b}{\left[1+\left(\frac{r}{b}\right)^2\right]^{3/2}} + \frac{1}{3}\frac{r/b}{\sqrt{1+\left(\frac{r}{b}\right)^2}} \right\} = \\ &= \frac{4}{3}\pi\rho_0 b^3 \frac{(r/b)^3}{\left[1+\left(\frac{r}{b}\right)^2\right]^{3/2}} \quad \Longrightarrow \quad M_{tot} = \lim_{r\to\infty} M(r) = \frac{4}{3}\pi\rho_0 b^3. \end{aligned} \tag{3.48}$$

Given the above expression for $M(r)$, the force per unit mass felt by a particle at distance r to the centre is

$$\mathbf{F}(r) = -\frac{GM_{tot}}{b^2}\frac{r/b}{\left[1+\left(\frac{r}{b}\right)^2\right]^{3/2}}\,\mathbf{e}_r, \tag{3.49}$$

which, near the centre ($r/b \ll 1$), at first order in r/b behaves linearly in r, as the force in a homogeneous sphere, while it has a Keplerian behaviour at large distances from the centre, $r/b \gg 1$.

By direct integration of $\mathbf{F}(r)$, or by a straightforward integration of the Poisson's equation, the potential of the Plummer sphere is easily obtained as

$$U(r) = \frac{GM_{tot}}{b} \frac{1}{\sqrt{1 + (r/b)^2}}. \tag{3.50}$$

Let us note that the above potential can be considered a *mollification* of the point mass potential generated by a point of mass M_{tot}, where the mollification length scale (called also *softening* length) is b. The substitution of the purely Keplerian interaction between two particles with this softened one avoids the *UV* divergence but, of course, it does not preserve the overall Newtonian behaviour of the system. Some further comments at this regard are found in Appendix A.

The potential energy of this distribution of matter is obtained, for instance, by Eq. 2.148, as follows

$$\begin{aligned} \Omega &= -\frac{1}{8\pi G} \iiint\limits_{\mathbb{R}^3} |\nabla U|^2 d^3\mathbf{r} = -\frac{1}{8\pi G} \int\limits_0^\infty \left[\frac{GM(r)}{r^2}\right]^2 4\pi r^2 \,\mathrm{d}r = \\ &= -\frac{1}{2}\frac{GM_{tot}^2}{b} \int\limits_0^\infty \frac{(r/b)^4}{\left[[1 + (r/b)^2\right]^3} \mathrm{d}\left(\frac{r}{b}\right) = -\frac{1}{2}\frac{GM_{tot}^2}{b} \int\limits_0^\infty \frac{x^4}{\left(1+x^2\right)^3} \mathrm{d}x. \end{aligned} \tag{3.51}$$

The improper integral in the rightmost side above converges because for $x \gg 1$ the integrand scales as $1/x^2$. The integral is solved by parts, letting $u = x^3$ and $dv = \dfrac{x}{\left(1+x^2\right)^3}$, leading to

$$\int\limits_0^\infty \frac{x^4}{\left(1+x^2\right)^3} \mathrm{d}x = \left[-\frac{1}{4}\frac{x^3}{\left(1+x^2\right)^2}\right]_0^\infty + \frac{3}{4}\int\limits_0^\infty \frac{x^2}{\left(1+x^2\right)^2} \mathrm{d}x = \frac{3}{4}\int\limits_0^\infty \frac{x^2}{\left(1+x^2\right)^2} \mathrm{d}x. \tag{3.52}$$

A further integration by parts of the integral in the rightmost hand side of Eq. 3.52 gives

$$\int\limits_0^\infty \frac{x^2}{\left(1+x^2\right)^2} \mathrm{d}x = -\left[\frac{1}{2}\frac{x}{1+x^2}\right]_0^\infty + \frac{1}{2}\int\limits_0^\infty \frac{\mathrm{d}x}{1+x^2} = \frac{1}{2}\int\limits_0^\infty \frac{\mathrm{d}x}{1+x^2}, \tag{3.53}$$

so that, finally

$$\int_0^\infty \frac{x^4}{\left(1+x^2\right)^3}\,\mathrm{d}x = \frac{3}{8}\int_0^\infty \frac{\mathrm{d}x}{\left(1+x^2\right)} = \frac{3}{8}\left[\arctan x\right]_0^\infty = \frac{3\pi}{16}, \tag{3.54}$$

which eventually yields the following expression for the potential energy:

$$\Omega = -\frac{3\pi}{32}\frac{GM_{tot}^2}{b} = -\frac{\pi^3 G\rho_0^2 b^5}{6}. \tag{3.55}$$

It is worth noting that the ratio between the above potential energy and that of a homogeneous sphere of constant density ρ_0 and radius $R = b$ (see Eq. 2.5) is $5\pi/32 \simeq 0.491$. Same value is obtained for the ratio computed at same total mass and $R = b$.

3.4.2 *The Isochrone Potential*

Another potential which, like the Plummer potential, is nearly harmonic in the central region and falls off as Keplerian at large distances from the centre, is the isochrone potential. It was introduced in 1959 by Michel Hénon [8] and its name derives from the fact that it is the most general spherical potential for which the radial oscillation period is independent of the angular momentum. Actually, this potential is expressed as

$$U(r) = \frac{GM}{b+\sqrt{b^2+r^2}} = \frac{GM}{b}\frac{1}{1+\sqrt{1+(r/b)^2}}, \tag{3.56}$$

characterized by 2 free parameters: the total mass, M, and the length scale, b. The corresponding force per unit mass is

$$\mathbf{F}(r) = \frac{\mathrm{d}U}{\mathrm{d}r}\mathbf{e}_r = -\frac{GM}{b^2}\frac{r/b}{\left(1+\sqrt{1+(r/b)^2}\right)^2\sqrt{1+(r/b)^2}}\mathbf{e}_r, \tag{3.57}$$

which, at first order in r/b, yields

$$\frac{\mathrm{d}U}{\mathrm{d}r} = \begin{cases} -\dfrac{GM}{b^2}\dfrac{1}{(r/b)^2} & r/b \gg 1, \\[2ex] -\dfrac{1}{4}\dfrac{GM}{b^2}\dfrac{r}{b} & r/b \ll 1. \end{cases} \tag{3.58}$$

Using Poisson's equation, it is easily seen that the density distribution which generates the potential 3.56 is

$$\rho(x) = \frac{M}{4\pi b^3}\frac{(2t+3)x^2 + 3(t+1)}{t^4(t+1)^3}, \tag{3.59}$$

where $x \equiv r/b$ and $t = t(x) \equiv \sqrt{1+x^2}$.

The above expression gives a finite central density $\rho(0) = 3M/(16\pi b^3)$.

Note that the asymptotic behaviours (for small and large distances from the centre) of the density are

$$\rho(x) \approx \begin{cases} \rho(0)\left(1 - \frac{5}{3}x^2\right), & x \ll 1, \\ \frac{M}{4\pi b^3}\left(\frac{2}{x^4} - \frac{3}{x^5}\right), & x \gg 1. \end{cases} \tag{3.60}$$

Note the shallower decrease of ρ at large distances compared to the Plummer case ($\rho \propto r^{-4}$ versus $\rho \propto r^{-5}$). The total mass

$$M = \lim_{x\to\infty} 4\pi b^3 \int_0^x \rho(x)x^2\,\mathrm{d}x, \tag{3.61}$$

is finite, because the integral convergence is guaranteed by the fact that, for $x \gg 1$, $\rho(x)x^2 \propto x^{-2}$.

The potential energy is

$$\begin{aligned} \Omega &= -\frac{1}{2}\int_0^\infty U(r)\rho(r)4\pi r^2\,\mathrm{d}r = \\ &= -\frac{1}{2}\frac{GM^2}{b}\int_0^\infty \frac{x^2}{1+\sqrt{1+x^2}} f(x,t(x))\,\mathrm{d}x \simeq -0.1187\frac{GM^2}{b}, \end{aligned} \tag{3.62}$$

where, in the above integral, $f(x,t(x))$ is the function that multiplies $M/(4\pi b^3)$ in Eq. 3.59. The integral in (3.62) is complicated, and we leave its explicit determination to the reader, as an exercise. We limit to note that the integral is convergent because for $x \gg 1$ the integrand behaves as $\rho(x)x \propto x^{-3}$.

Note that the ratio between the above potential energy and that of a Plummer sphere of same mass and length scale b is $\Omega_{is}/\Omega_P \simeq 0.4032$. The potential energy ratio at same $\rho(0)$ and length scale is $\Omega_{is}/\Omega_P \simeq 6.448$.

3.5 Quality of Motion

Although the spherical symmetry represents a significant simplification, the integration of the equation of motion remains a task that cannot, in general, be faced on an analytical basis but requires a numerical approach. Anyway, when the density, and so the potential, are spherically symmetric, a *qualitative* approach to the problem is possible, which helps because it restricts the phase space allowed to the motion.

The qualitative treatment is based on the constants and integrals of motion. The constants and integrals of motions, indeed, pose constraints to regions available to the actual motion. In the case of spherical potentials, we recall that both the specific mechanical energy, E, and the specific angular momentum, $\mathbf{L}$, are integrals of motion.

The mechanical energy per unit mass of a particle in a spherical potential $U(r) \equiv -V(r)$[2] is (in spherical polar coordinates)

$$E = \frac{1}{2}(\dot{r}^2 + r^2 \sin^2 \varphi \dot{\theta}^2 + r^2 \dot{\varphi}^2) + V(r). \tag{3.63}$$

We saw in Sect. 3.2 that in a spherical force field the motion is planar, so that, choosing polar coordinates such that the equatorial plane ($\varphi = \pi/2$) is the motion plane, the expression above simplifies in

$$E = \frac{1}{2}(\dot{r}^2 + r^2\dot{\theta}^2) + V(r) = \frac{1}{2}\dot{r}^2 + \frac{1}{2}\frac{L^2}{r^2} + V(r), \tag{3.64}$$

so that, for the motion to be physically possible, we require

$$\frac{1}{2}\dot{r}^2 = E - \frac{1}{2}\frac{L^2}{r^2} - V(r) = E - V_{eff}(r; L^2) \geq 0, \tag{3.65}$$

i.e. $E \geq V_{eff}(r; L^2)$, where the function $V_{eff}(r; L^2) \equiv -U_{eff}(r; L^2) = \frac{1}{2}(L^2/r^2) + V(r)$ is called *effective* potential energy.

The condition above constitutes an actual limitation for the physical portion of plane allowed to the motion of the particle of given E and L^2 as defined by the initial conditions.

3.5.1 The Keplerian Case

The Keplerian, point mass, potential is the simplest to approach and so it is the archetype of the *qualitative* analysis of motion. Whenever the test particle mass, m, is much smaller than the other mass, M, i.e. $m \ll M$, the problem reduces to a *one*

[2] In our notation, $U(r)$ is the potential and $V(r)$ is the potential energy.

body problem, because the centre of mass practically coincides with the position of the massive particle. In any case, also when m/M is not much smaller than one, qualitative considerations hold for the relative motion, i.e. the motion of one of the two bodies around the other.

The effective potential energy of the test object is

$$V_{eff}(r; L^2) = \frac{1}{2}\frac{L^2}{r^2} - \frac{GM}{r}, \tag{3.66}$$

and, as we said above, the motion is possible whenever the test particle energy is $E \geq V_{eff,min}$, where $V_{eff,min}$ is the minimum in r of $V_{eff}(r; L^2)$ for given L^2. In the case of $L^2 = 0$, the motion is possible (and radial) for every energy E, because (see Fig. 3.3) $\lim_{r\to 0} V_{eff}(r; 0) = -\infty$, while, for $L^2 > 0$, $V_{eff}(r; L^2)$ takes its minimum value, $V_{eff,min} = -\frac{1}{2}(GM/L)^2$, at $r_{min} = L^2/(GM)$.

Clearly, the function $V_{eff}(r; L^2)$ is continuous for every r but the origin, which is singular for both $L^2 = 0$, where V_{eff} diverges to $-\infty$, and $L^2 > 0$ where it diverges positively (*centrifugal barrier*).

For the sake of clarity we can distinguish the *non-radial* case ($L^2 > 0$) from the *radial* one ($L^2 = 0$).

Case $L^2 > 0$ Qualitatively speaking, this case corresponds to two of the example curves in Fig. 3.3. The upper curve corresponds to a value of L^2 twice that of the other curve with $L^2 > 0$. The stationary values of $r(t)$ are solutions of the equation $\dot{r} = \pm\sqrt{2\left[E - V_{eff}(r; L^2)\right]} = 0$, which, for $E \neq 0$, correspond to the roots of the quadratic equation

$$Er^2 + GMr - \frac{1}{2}L^2 = 0, \tag{3.67}$$

which are

$$r_{1,2} = \frac{-GM \pm \sqrt{(GM)^2 + 2EL^2}}{2E}. \tag{3.68}$$

These solutions are:

(i) complex conjugates of each other when $(GM)^2 + 2EL^2 < 0$.
Actually, this corresponds to $E < -\frac{1}{2}(\frac{GM}{L})^2 = V_{eff,min}$, and means *no motion*, as we said before;

(ii) two real and coincident when $(GM)^2 + 2EL^2 = 0$.
This means that for $E = V_{eff,min}$, the pericentre, r_-, and the apocentre, r_+, distances are equal, so that the motion is along a circle (circular orbit);

(iii) two real and distinct when $(GM)^2 + 2EL^2 > 0$.
If $E > 0$, the solution with the $-$ sign in front of the square root in (3.68) is negative and must be excluded while the one with $+$ is positive

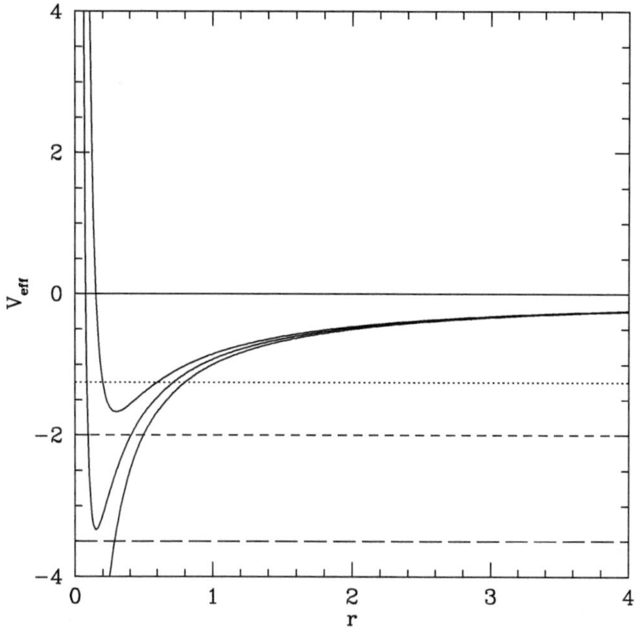

Fig. 3.3 Behaviour of $V_{eff}(r; L^2)$ (in arbitrary units) in the Keplerian case, for $L^2 = 0$ (lower curve) and for two non-zero values of L^2. The dotted, short-dashed and long-dashed horizontal lines refer to examples of level of energies for which oscillations are radially limited for the three choices of L^2

and corresponds to the pericentre distance, r_-. The trajectory is unbound (hyperbolic).

If $E < 0$ the solutions are both positive and r_- corresponds to taking the $+$ sign and r_+ to the $-$ sign. The trajectory is bound (elliptic).

If $E = 0$, the equation for stationary points of $r(t)$ reduces to $GMr - L^2/2 = 0$, and so there is only one stationary point, $r_- = \frac{1}{2}L^2/(GM)$, and the trajectory is unbound (parabolic).

Case $L^2 = 0$
This case corresponds to the lowest curve in Fig. 3.3, which plots the function $V_{eff}(r; 0) = -(GM)/r$, which goes to $-\infty$ for $r \to 0$ and is always negative. For $L^2 = 0$ Eq. 3.67 to determine extremes of radial motion reduces to the equation $r(Er + GM) = 0$, whose solutions are $r_1 = r_- = 0$ and $r_2 = -(GM)/E$. This latter solution is positive for $E < 0$, which gives the apocentre distance r_+, while, for $E > 0$, it is negative, meaning that for $E > 0$ the motion is unbound, like in the case $L^2 > 0$.

If $E = 0$ the only solution is $r_1 = r_- = 0$ and the motion, as in the case $L^2 > 0$, is unbound.

3.5.2 The Homogeneous Sphere Case

In the case of a homogeneous sphere of mass density ρ_0, we have (see Eq. 3.27)

$$V_{eff}(r; L^2) = \begin{cases} \frac{1}{2}\frac{L^2}{r^2} + 2\pi G\rho_0 \left(\frac{1}{3}r^2 - R^2\right), & \text{if } r \le R, \\ \frac{1}{2}\frac{L^2}{r^2} - \frac{4}{3}\pi G\rho_0 \frac{R^3}{r}, & \text{if } r > R. \end{cases} \tag{3.69}$$

We can, again, distinguish the non-radial case ($L^2 > 0$) from the radial one ($L^2 = 0$).

Case $L^2 > 0$
From Eq. 3.69 it is

$$\lim_{r\to 0} V_{eff}(r; L^2) = +\infty, \qquad \lim_{r\to\infty} V_{eff}(r; L^2) = 0^-, \tag{3.70}$$

conditions that, by continuity of $V_{eff}(r; L^2)$ for $r > 0$, imply that $V_{eff}(r; L^2)$ has either only one stationary point or it has an odd number of stationary points. In this specific case (Eq. 3.70), we have[3]

$$V'_{eff} = \begin{cases} -\frac{L^2}{r^3} + \frac{4}{3}\pi G\rho_0 r, & \text{if } r \le R, \\ -\frac{L^2}{r^3} + \frac{4}{3}\pi G\rho_0 \frac{R^3}{r^2}, & \text{if } r > R, \end{cases} \tag{3.71}$$

so that

$$V'_{eff} = 0 \Longleftrightarrow \begin{cases} r_1 = \sqrt[4]{L^2 \frac{3}{4\pi G\rho_0}}, \\ r_2 = L^2 \frac{3}{4\pi G\rho_0 R^3}. \end{cases} \tag{3.72}$$

The above zeros of the $V_{eff}(r; L)$ radial derivative have sense if $r_1 \le R$ and $r_2 \ge R$, which reflect into conditions for L^2:

$$\begin{cases} r_1 \le R \Longleftrightarrow & L^2 \le \frac{4}{3}\pi G\rho_0 R^4, \\ r_2 \ge R \Longleftrightarrow & L^2 \ge \frac{4}{3}\pi G\rho_0 R^4. \end{cases} \tag{3.73}$$

[3] Hereafter, we indicate by apices, $'$, derivatives with respect to r.

Naturally, a lower angular momentum results in a more limited radial oscillation compared to a higher angular momentum. The above threshold value for L^2 corresponds to the squared angular momentum of the circular orbit of radius R, $L_c(R) = R^2\sqrt{(4\pi G\rho_0)/3}$.

Summarizing, if $L^2 \leq L_c^2(R)$ then r_1 is the stationary point, otherwise the stationary point is r_2.

Actually, the stationary values r_1 and r_2 are minima for $V_{eff}(r; L^2)$, because

$$V''_{eff}(r_1; L^2) = \frac{16\pi}{3} G\rho_0 > 0, \tag{3.74}$$

and

$$V''_{eff}(r_2; L^2) = \left(\frac{4\pi}{3} G\rho_0 \frac{R^3}{L^{3/2}}\right)^4 > 0. \tag{3.75}$$

Moreover

$$V_{eff}(r_1; L^2) = L\sqrt{\frac{4\pi}{3} G\rho_0} - 2\pi G\rho_0 R^2, \tag{3.76}$$

so that the condition $L^2 \leq L_c^2(R)$ implies

$$V_{eff}(r_1; L^2) \leq L_c(R)\sqrt{\frac{4\pi}{3} G\rho_0} - 2\pi G\rho_0 R^2 = -\frac{2\pi}{3} G\rho_0 R^2 < 0. \tag{3.77}$$

On the other hand, for r_2 it is

$$V_{eff}(r_2; L^2) = -\frac{1}{2}\left(\frac{4\pi}{3} G\rho_0 \frac{R^3}{L}\right)^2 < 0.$$

In conclusion, for any value of L, radially periodic motion is possible only if $E < 0$.

It is worth noting that for $L^2 < L_c^2(R)$ the trajectories always penetrate the sphere ($r_- < R, \ \forall E$), while for $L^2 > L_c^2(R)$ it is $r_+ > R, \ \forall E$.

Note also that a sufficient (but not *necessary*) condition for radially limited orbits to be all external to the sphere is $L^2 > 2L_c^2(R)$, for any value of $V_{eff,min} < E < 0$. This is because for $L^2 = 2L_c^2(R)$ we have $V_{eff}(R; L^2) = 0$.

Summarizing:

$$\begin{aligned}
&\text{(a) } E < V_{eff,min}: && \text{no motion;} \\
&\text{(b) } E = V_{eff,min}: && \text{circular motion,} \quad r = r_- = r_+, \ \forall t; \\
&\text{(c) } V_{eff,min} < E < 0: && \text{radially bound motion,} \quad r_- \leq r \leq r_+, \ \forall t; \\
&\text{(d) } E \geq 0: && \text{radially unbound motion;} \quad r_+ = \infty.
\end{aligned} \tag{3.78}$$

Note that case (d) is that of a particle moving faster than the escape velocity, since $E \geq 0$ corresponds to

$$\frac{1}{2}v^2 \geq U(r) \Longleftrightarrow v \geq \sqrt{2U(r)} \equiv v_{esc}(r). \tag{3.79}$$

The pericentre, $r_-(E, L^2)$, and apocentre, $r_+(E, L^2)$, distances are stationary points for $r(t)$ and, so, they are the roots of the equation

$$\dot{r}^2 = 2\Big[E - V_{eff}(r; L^2)\Big] = 0. \tag{3.80}$$

We leave to the reader the discussion of the existence of those roots and their dependence upon E and L^2.

Case $L^2 = 0$
As we know, this case corresponds to a radial motion. If ${}^2L = 0$, it results $V_{eff}(r; L^2) = V(r)$ and, in the specific case of a homogeneous sphere, $V(0) = -2\pi G\rho_0 R^2 < 0$, and $V'(r) = 0$ both for $r = 0$ and $r \to \infty$, moreover

$$V''(r) = \begin{cases} \dfrac{4}{3}\pi G\rho_0 > 0, & r \leq R \\ -\dfrac{8}{3}\pi G\rho_0 \left(\dfrac{R}{r}\right)^3 < 0, & r > R. \end{cases} \tag{3.81}$$

The expressions above say that the minimum is attained in $r = 0$, which is so a stable equilibrium point.

The qualitative characteristics of the $L^2 = 0$ motion are summarized as:

$$\begin{aligned}
&(a)\ E < -2\pi G\rho_0 R^2, \quad \text{no motion;} \\
&(b)\ E = -2\pi G\rho_0 R^2, \quad \text{particle sitting at the origin: } r = 0; \\
&(c)\ -2\pi G\rho_0 R^2 < E < 0, \quad \text{radial oscillation through the origin: } 0 \leq r \leq r_+; \\
&(d)\ E = 0, \quad \text{asymptotically unbound motion: } r_+ = \infty; \\
&(e)\ E > 0, \quad \text{unbound motion: } r_+ = \infty.
\end{aligned} \tag{3.82}$$

Recalling Sect. 3.3.2, in the case $L^2 = 0$ the dynamical equation for $r(t)$ (Eq. 3.41) reduces to that of a 3D harmonic oscillator for $E \leq V(R)$ (the whole oscillation is internal to the sphere). If $E > V(R)$ the oscillation is harmonic for $r \leq R$, matched to a Keplerian motion for $r > R$. Actually, for $L^2 = 0$ the equation of motion is

$$\ddot{r} = \frac{\mathrm{d}U}{\mathrm{d}r} = \begin{cases} -\frac{4}{3}\pi G\rho_0 r, & r \leq R, \\ -\frac{4}{3}\pi G\rho_0 \frac{R^3}{r^2}, & r \geq R. \end{cases} \tag{3.83}$$

Note 12
As we saw, in a spherical potential, which is Keplerian out of the sphere boundary R, the condition for a particle to be unbound is $E \geq 0$, which is equivalent to $v \geq v_{esc}(r) \equiv \sqrt{2U(r)}$. On this purely classical basis, John Michell in 1783 deduced that a sufficiently massive planet could have been completely dark because even light would not have been fast enough to escape the gravitational attraction.

Under the hypothesis of a planet of uniform mass density, ρ_0, the condition for the planet to be dark would be that the escape speed at its surface $v_{esc}(R) = R\sqrt{8\pi G\rho_0/3}$ is greater than the speed of light in vacuum, c. This means, for a planet with the average density of the Earth (5.51 g cm^{-3}), to have a radius $R \geq 1.7 \times 10^8$ km $\simeq 243\mathrm{R}_\odot \simeq 1.13$ AU! Although historically interesting, this implicit deduction of the existence of an *event horizon* is physically flawed, because it bases on the wrong assumption that the light behaves as "particles" with a non-zero finite mass, so as to allow writing their mechanical energy per unit mass in the usual way.

Exercise 3.4 Compute, for $L^2 = 0$, the pericentre and apocentre distances for oscillation within the homogeneous sphere.

Solution

$\dot{r} = 0 \Longleftrightarrow E - V_{eff}(r) = 0$, i.e.

$$E - 2\pi G\rho_0 \left(\frac{r^2}{3} - R^2\right) = 0,$$

$$-\frac{2}{3}\pi G\rho_0 r^2 + E + 2\pi G\rho_0 R^2 = 0 \quad \Longrightarrow r_+ = \sqrt{3R^2 + \frac{E}{(2/3)\pi G\rho_0}}.$$

The condition for the oscillation to remained confined within the sphere is $r_+ \leq R$, which reflects into a condition for the energy. $E \leq -4\pi G\rho_0 R^3/3$.

As expected, there is only one solution acceptable for the quadratic equation (positive determination of the square root). The pericentre $r_- = 0$ is not a point where $\dot{r}$ vanishes; rather, at the time $\bar{t}$ of crossing the origin, $r(t)$ is not differentiable (the radial velocity changes sign discontinuously).

3.6 Periods of Oscillations

As seen above, in a spherical potential, once given the (constant) values of E and L^2 the motion is allowed in the radial region where $E \geq V_{eff}(r; L^2)$. The motion can either be radially bound, meaning the existence of a pericentre distance r_- and a finite apocentre distance r_+, or unbound, this latter corresponding to $r_+ = +\infty$. Both r_- and r_+ depend, in general, on E and L^2.

The number of different radial intervals of oscillation, and so the number of pairs (r_-, r_+), depends upon the shape of the potential $U(r)$, because for any minimum of $V_{eff}(r; L^2)$ there is always a local radial oscillation occurring in a finite time for energies larger than the local minimum of $V_{eff}(r; L^2)$ and lower than the lowest closer maximum of $V_{eff}(r; L^2)$ (see Figs. 3.4 and 3.5).

The time needed to make a complete radial oscillation between r_- and r_+ (called *radial period*, $T_r(E, L^2)$, as already said in Sect. 3.3.2) is formally given by

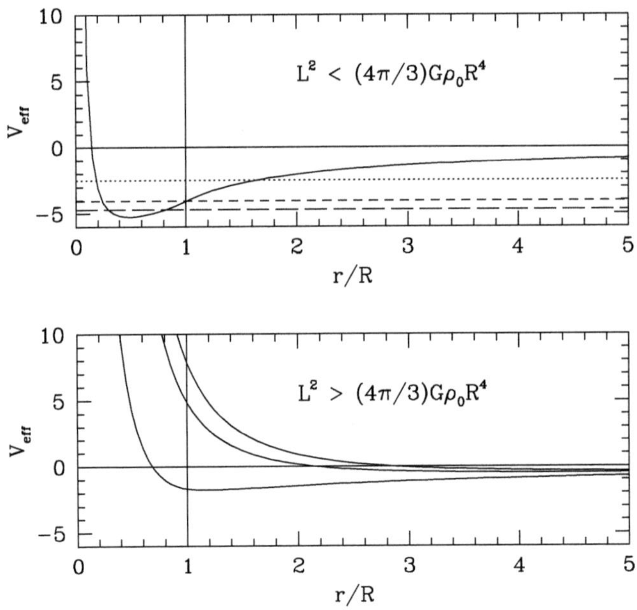

Fig. 3.4 Behaviour of $V_{eff}(r; L^2)$ vs radius for a homogeneous sphere, in the two regimes of low (upper panel) and high (lower panel) specific orbital angular momentum. In both panels, the horizontal solid line refers to zero energy and the horizontal solid line to the sphere radius location. In the upper panel the straight lines at decreasing energies corresponds to cases of: (i) oscillation through the sphere boundary (dotted line); (ii) oscillations touching internally the sphere boundary (short dashed line); (iii) oscillations all within the sphere (long dashed line). The lower panel reports (from left to right) plots for three different increasing values of L^2. V_{eff} is in arbitrary units

$$T_r = \int_{r_-}^{r_+} \frac{dr}{\dot{r}} + \int_{r_+}^{r_-} \frac{dr}{\dot{r}} = \int_{r_-}^{r_+} \frac{dr}{\sqrt{2(E - V_{eff})}} + \int_{r_+}^{r_-} \frac{dr}{-\sqrt{2(E - V_{eff})}} = \\ = 2\int_{r_-}^{r_+} \frac{dr}{\sqrt{2(E - V_{eff})}}. \tag{3.84}$$

This characteristic time involves an integral with singularity in both r_- and r_+, where the denominator vanishes, and can be finite or infinite in dependence on convergence or divergence of the improper integral.

We examine the two representative cases:

$$(i)\ V'_{eff}(r_-; L^2) < 0 \text{ and } V'_{eff}(r_+; L^2) > 0;$$
$$(ii)\ V'_{eff}(r_-; L^2) = 0, \text{or } V'_{eff}(r_+; L^2) = 0, \text{ or they are both equal to 0.} \tag{3.85}$$

Case (i)

In this case the integrals at the rhs of Eq. 3.84 converge.

Actually, possible divergence could come just by behaviour of the integrand around r_- and/or r_+, so we expand $V_{eff}(r; L^2)$ around r_- and r_+, to get

$$V_{eff}(r; L^2) = V_{eff}(r_\pm; L^2) + V'_{eff}(\xi_\pm; L^2)(r - r_\pm), \tag{3.86}$$

synthesizing in a single formula the two expansions around r_- and r_+, with clear meaning of the symbol $r_\pm$. Note that $\xi_\pm$ is an unknown point in the neighbourhood of r_- (if we study convergence around r_-, so we call it ξ_-) or r_+ (if we study convergence around r_+, so we call it ξ_+). Note, also, that by continuity $V'_{eff}(\xi_-; L^2) < 0$ and $V'_{eff}(\xi_+; L^2) > 0$. This leads to

$$\frac{1}{\sqrt{2(E - V_{eff})}} = \frac{1}{\sqrt{2\left|V'_{eff}(\xi_\pm; L^2)(r - r_\pm)\right|}}, \tag{3.87}$$

because $E = V_{eff}(r_\pm; L^2)$, leading to that the integral on the rightmost hand side of Eq. 3.84 is convergent both in r_- and r_+. Let us prove convergence around r_- (around r_+ the demonstration is identical). Around r_-, the integral under study is

$$\int_{r_-}^{r_- + \epsilon} \frac{dr}{\sqrt{2(E - V_{eff})}}, \tag{3.88}$$

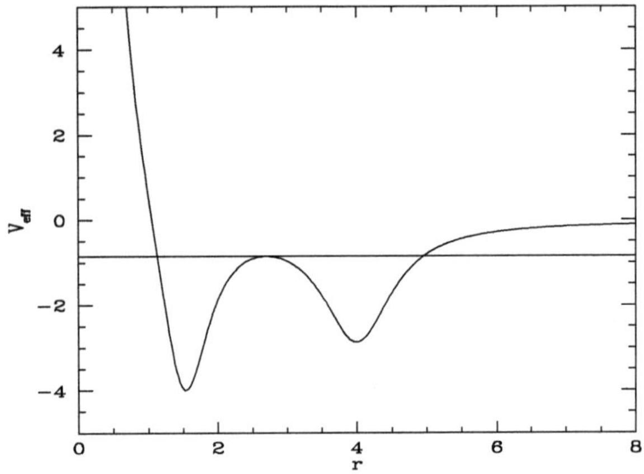

Fig. 3.5 Sketch of an effective potential (in arbitrary units) with centrifugal barrier in $r = 0$ and having two wells. The horizontal line refers to the energy required to a particle to reach the relative maximum separating the two wells

which, at first order in $\epsilon > 0$ is (see Eq. 3.87)

$$\int_{r_-}^{r_-+\epsilon} \frac{\mathrm{d}r}{\sqrt{2(E - V_{eff})}} = \sqrt{\frac{2}{|V'_{eff}(\xi_-; L)|}} \int_{r_-}^{r_-+\epsilon} \frac{\mathrm{d}r}{r - r_-} = 2\sqrt{\frac{2}{|V'_{eff}(\xi_-; L)|}}\sqrt{\epsilon}, \tag{3.89}$$

which vanishes as $\epsilon \to 0$, guaranteeing the convergence of the integral in Eq. 3.84 in the lower integration limit. By the same token, the integral converges also at the upper limit.

Case (ii)
Let us consider $V'_{eff}(r_-; L^2) = 0$ (the case $V'_{eff}(r_+; L^2) = 0$ is equivalent). Figure 3.5 displays an example of such a situation (where the local maximum corresponds to apocentre distance, r_+, if the initial distance is leftward or to pericentre distance, r_-, if the initial distance is rightward).

The lowest-order approximation of $V_{eff}(r; L^2)$ is, for r in a neighbourhood of r_- where $V''_{eff} < 0$,

$$V_{eff}(r; L^2) = V_{eff}(r_-; L^2) + \frac{1}{2}V''_{eff}(\eta_-; L^2)(r - r_-)^2, \tag{3.90}$$

where η_- is an unknown point in the right neighbourhood of r_-.

Consequently

$$\frac{1}{\sqrt{2(E - V_{eff})}} = \frac{1}{\sqrt{-V''_{eff}(\eta_-; L)(r - r_-)^2}} = \frac{1}{\sqrt{|V''_{eff}(\eta_-; L)|}}\frac{1}{r - r_-}, \tag{3.91}$$

so that

$$T_r(E, L) = \frac{2}{\sqrt{|V''_{eff}(\eta_-; L)|}} \int_{r_-}^{r_+} \frac{dr}{r - r_-}, \tag{3.92}$$

which diverges logarithmically in the lower integration limit.

Note 13
In the spherical gravitational case, the existence of more than one potential well is excluded. Actually, a density $\rho(r)$ generates a gravitational force whose absolute value is $GM(r)/r^2$. The centrifugal equilibrium condition is (with v_θ tangential velocity)

$$\frac{v_\theta^2}{r} = r\dot{\theta}^2 = G\frac{M(r)}{r^2},$$

which, by elimination of $\dot{\theta}$ by the integral of angular momentum $L = r^2\dot{\theta} = const.$, becomes

$$L^2 = GM(r)\, r.$$

For any chosen value of L^2, the above equation has a unique solution because the function $M(r)\, r$ is a continuous and strictly increasing function of r that is zero at the origin and diverges as $r \to +\infty$. This means that, for any given value of the angular momentum, there is only one circular orbit, which is characterized, indeed, by radius r_c such that $V'_{eff}(r_c; L^2) = 0$. It is known that necessary and sufficient condition for the linear stability is that $V''_{eff}(r; L^2) > 0$, which corresponds to having finite values for r_- and r_+ for the radial oscillation around r_c.

Exercise 3.5 Show that among all possible orbits of given energy E in a gravitational spherical potential $U(r)$, the circular orbit is that of maximum angular momentum.

Solution Given the spherical potential $U(r)$, and choosing polar coordinates in the xy-plane of motion (with z-axis oriented as $\mathbf{L}$), the energy per unit mass is

$$E = \frac{1}{2}\left(\dot{r}^2 + r^2\dot{\theta}^2\right) - U(r) = \frac{1}{2}\dot{r}^2 + \frac{1}{2}\frac{L^2}{r^2} - U(r),$$

so that (being $\dot{r}^2 \geq 0$)

$$L^2 = 2r^2\left[E - \frac{1}{2}\dot{r}^2 + U(r)\right] \leq 2r^2\left[E + U(r)\right],$$

for every chosen r. Letting $g(r) = 2r^2[E + U(r)]$, we have to show that the circular orbit radius, r_c, maximizes the function $g(r)$ for a fixed E. To find critical points of $g(r)$ we seek for solutions of the equation:

$$g'(r) = \frac{\mathrm{d}}{\mathrm{d}r}\left\{2r^2[E + U(r)]\right\} = 4r[E + U(r)] + 2r^2U'(r) = 0. \tag{3.93}$$

Now, for a circular orbit of energy E it is

$$E = \frac{1}{2}\frac{L_c^2}{r_c^2} - U(r_c),$$

and, from Eq. 3.93, $L_c^2 = -r_c^3U'(r_c)$, so that we can write

$$E + U(r_c) = -\frac{1}{2}r_cU'(r_c).$$

Substituting $E+U(r_c)$ from the equation above into the condition 3.26 for critical points evaluated at r_c it results

$$2\left[-\frac{1}{2}r_cU'(r_c)\right] + r_cU'(r_c) = -r_cU'(r_c) + r_cU'(r_c) = 0,$$

that confirms r_c as a stationary point for $g(r)$. It is actually a maximum, because $g''(r_c) = 6r_cU'(r_c) + 2r_c^2U'' < 0$ (the proof, based on stability of circular orbits in gravitational spherical potentials, is left to the reader).

To conclude, because $g(r_c) = L_c^2$, we have finally that for every r along the particle path

$$L^2 \le g(r) \le g(r_c) = L_c^2.$$

3.6.1 Radial Period in the Keplerian Potential

In the Keplerian potential generated by a point of mass M, the period of radial oscillations is written as

$$T_r = 2\int_{r_-}^{r_+} \frac{\mathrm{d}r}{\sqrt{2\left(E - \frac{1}{2}\frac{L^2}{r^2} + \frac{GM}{r}\right)}} = \sqrt{2}\int_{r_-}^{r_+} \frac{r\,\mathrm{d}r}{\sqrt{Er^2 - \frac{1}{2}L^2 + GMr}}, \tag{3.94}$$

where, as usual, $r_\pm$ are the minimum and maximum radial distances, solutions of

$$Er^2 + GMr - \frac{1}{2}L^2 = 0. \tag{3.95}$$

As the qualitative study of the Kepler potential taught us (see Sect. 3.5.1), the motion is *bound* (r_+ is finite) or *unbound* ($r_+ = \infty$) when $V_{eff,min} \leq E < 0$ or when $E \geq 0$, respectively, independently of L.

We remind that, when $L^2 = 0$, then $r_- = 0$. The dependence of $r_\pm$ on E when $L^2 > 0$ has been discussed in (3.5.1).

Due to that radial periodicity (r_+ finite) requires $E < 0$ and $E > V_{eff,min}$, and consequently $(GM)^2 + 2EL^2 > 0$, the rightmost integral in Eq. 3.94 can be rewritten as follows, after letting $a \equiv E < 0$, $b \equiv GM > 0$, and $c \equiv -L^2/ < 0$:

$$T_r = \sqrt{2} \int_{r_-}^{r_+} \frac{r\mathrm{d}r}{\sqrt{ar^2 + br + c}}, \tag{3.96}$$

which can be solved by parts, with the choices $u = r$ and $\mathrm{d}v = \mathrm{d}r/\sqrt{ar^2 + br + c}$, so to have

$$T_r = \sqrt{2} \left\{ \left[-\frac{r}{\sqrt{-a}} \arcsin\left(\frac{2ar + b}{\sqrt{\Delta}}\right) \right]_{r_-}^{r_+} + \frac{1}{\sqrt{-a}} \int_{r_-}^{r_+} \arcsin\left(\frac{2ar + b}{\sqrt{\Delta}}\right) \mathrm{d}r \right\}, \tag{3.97}$$

where $\Delta = b^2 - 4ac = (GM)^2 + 2EL^2 > 0$.

Now,

$$T_r = \sqrt{2} \left\{ \left[-\frac{r}{\sqrt{-a}} \arcsin\left(\frac{2ar + b}{\sqrt{\Delta}}\right) \right]_{r_-}^{r_+} + \frac{\sqrt{\Delta}}{\sqrt{-a}} \frac{1}{2a} \int_{t_-}^{t_+} \arcsin t \, \mathrm{d}t \right\}, \tag{3.98}$$

where we introduced $t \equiv \dfrac{2ar + b}{\sqrt{\Delta}}$, which implies $t_- = \dfrac{2ar_- + b}{\sqrt{\Delta}}$ and $t_+ = \dfrac{2ar_+ + b}{\sqrt{\Delta}}$.

The integral $\int \arcsin t \, \mathrm{d}t$ in Eq. 3.98 is, also, solvable by parts ($u = \arcsin t$ and $\mathrm{d}v = \mathrm{d}t$) obtaining

$$\int \arcsin t \, \mathrm{d}t = t \arcsin t + \sqrt{1 - t^2} + c, \tag{3.99}$$

to give, finally

$$T_r = \sqrt{2} \left[-\frac{1}{\sqrt{-a}} \arcsin t \left(r - \frac{\sqrt{\Delta}}{2a} t \right) + \frac{1}{2a} \frac{\sqrt{\Delta}}{\sqrt{-a}} \sqrt{1 - t^2} \right]_{r_-}^{r_+} =$$

$$= \sqrt{2} \left[-\frac{1}{\sqrt{-a}} \arcsin(-1) \left(-\frac{b}{2a} \right) + \frac{1}{\sqrt{-a}} \arcsin(1) \left(-\frac{b}{2a} \right) \right] =$$

$$= \sqrt{2}\left[2\frac{1}{\sqrt{-a}}\left(-\frac{b}{2a}\right)\frac{\pi}{2}\right] = \frac{\sqrt{2}}{2}\pi\frac{b}{(-a)^{\frac{3}{2}}} = \frac{\sqrt{2}}{2}\pi\frac{GM}{(-E)^{\frac{3}{2}}}. \tag{3.100}$$

Note that T_r depends only on E and not on L.

3.6.2 Radial Period in the Homogeneous Sphere Potential

In any spatially limited spherical distribution of density ($\rho(r) > 0$ for $r \leq R$ with $R > 0$) the possible cases, for bound motion, in dependence on E and L, are:

$$\begin{aligned}
&(i)\ 0 \leq r_- < r_+ \leq R, && \text{oscillation } \mathit{within} \text{ thesphere;} \\
&(ii)\ 0 \leq r_- < R \leq r_+, && \text{oscillation } \mathit{in} \text{ and } \mathit{out} \text{ thesphere;} \\
&(iii)\ R \leq r_- < r_+, && \text{oscillation } \mathit{outside} \text{ thesphere.}
\end{aligned} \tag{3.101}$$

In the 3 cases we have:

$$\begin{aligned}
(i)\quad & T_r = 2\int_{r_-}^{r_+}\frac{\mathrm{d}r}{\sqrt{2(E - V_{eff}^{(int)})}}, \\
(ii)\quad & T_r = 2\left[\int_{r_-}^{R}\frac{\mathrm{d}r}{\sqrt{2(E - V_{eff}^{(int)})}} + \int_{R}^{r_+}\frac{\mathrm{d}r}{\sqrt{2(E - V_{eff}^{(ext)})}}\right], \\
(iii)\quad & T_r = 2\int_{r_-}^{r_+}\frac{\mathrm{d}r}{\sqrt{2(E - V_{eff}^{(ext)})}},
\end{aligned} \tag{3.102}$$

where $V_{eff}^{(int)}(r; L)$ and $V_{eff}^{(ext)}(r; L)$ are the expressions of the effective potential energy inside and out of the sphere, respectively. Let us discuss the specific case of the homogeneous sphere, $\rho(r) = \rho_0 > 0$, for $r \leq R$ and $\rho = 0$ for $r > R$.

Let us treat extensively case (i).

Recalling the expression in the top line of (3.69) for $V_{eff}^{(int)}$, we have

$$T_r = \sqrt{2}\int_{r_-}^{r_+}\frac{r\,\mathrm{d}r}{\sqrt{Ar^4 + Br^2 + C}}, \tag{3.103}$$

with

$$\begin{cases} A = -\dfrac{2}{3}\pi G\rho_0 < 0, \\ B = E + 2\pi G\rho_0 R^2 \geq 0, \\ C = -\dfrac{1}{2}L^2 \leq 0. \end{cases} \tag{3.104}$$

Note that $B \geq 0$ because, to have real motion, $E \geq V_{eff,min}(L) = L\sqrt{\frac{4\pi}{3}G\rho_0} - 2\pi G\rho_0 R^2$, as from Eq. 3.76, and, consequently, $B \geq L\sqrt{\frac{4\pi}{3}G\rho_0} > 0$.

For sake of clarity let distinguish the two cases, a) $L^2 = 0$, and b) $L^2 > 0$.

(a) If $L^2 = 0$ (purely radial motion, $r_- = 0$) the integral in Eq. 3.103 simplifies, because $C = 0$, to give

$$\begin{aligned} T_r &= \sqrt{2}\int_0^{r_+} \frac{\mathrm{d}r}{\sqrt{Ar^2 + B}} = \frac{\sqrt{2}}{\sqrt{B}}\int_0^{r_+} \frac{\mathrm{d}r}{\sqrt{1 + \dfrac{A}{B}r^2}} = \\ &= \frac{\sqrt{2}}{\sqrt{B}}\int_0^{y_+} \frac{\sqrt{-B/A}}{\sqrt{1-y^2}}\mathrm{d}y = \frac{\sqrt{2}}{\sqrt{-A}}\left[\arcsin y\right]_0^{y_+}, \end{aligned} \tag{3.105}$$

after setting $-\frac{A}{B}r^2 = y^2$ so as $y = \sqrt{-\frac{A}{B}}\,r$.

From Exercise 3.4, it results $y_+ = \sqrt{-A/B}r_+ = 1$, so that

$$T_r = \frac{\sqrt{2}}{\sqrt{\dfrac{2}{3}\pi G\rho_0}}\left[\arcsin y\right]_0^1 = \frac{\pi}{2}\frac{1}{\sqrt{\dfrac{\pi}{3}G\rho_0}}. \tag{3.106}$$

Note 14
The above result could have also been obtained taking into account that in a homogeneous sphere when $L^2 = 0$ the motion is radial and harmonic with pulsation $\omega^2 = \frac{4}{3}\pi G\rho_0$, which gives the motion period $T = 2\pi/\omega = \pi/\sqrt{(\pi G\rho_0)/3}$. By its definition, the radial period is $T_r = T/2$.

Let us now consider the generic $L^2 > 0$ case. In this case we have

$$T_r = \frac{\sqrt{2}}{2}\int_{r_-^2}^{r_+^2} \frac{\mathrm{d}r^2}{\sqrt{Ar^4 + Br^2 + C}} = \frac{\sqrt{2}}{2}\int_{x_-}^{x_+} \frac{\mathrm{d}x}{\sqrt{Ax^2 + Bx + C}}, \tag{3.107}$$

with the substitution $x = r^2$.

Given that $A < 0$, and $\Delta \equiv B^2 - 4AC = (E + 2\pi G\rho_0 R^2)^2 - \frac{4}{3}\pi G\rho_0 L^2 > 0$ because of the condition on $E > V_{eff,min}(L^2)$ for real motion (see Eq. 3.76), the integral in Eq. 3.107 can be solved thanks to the substitution $\sqrt{Ax^2 + Bx + C} = t(x - \alpha)$, where α is one of the two (real and distinct) roots of the radicand x_- and x_+ ($x_+ > x_-$), giving

$$T_r = -\frac{\sqrt{2}}{2}\frac{1}{\sqrt{-A}}\left[\arcsin\left(\frac{2Ax + B}{\sqrt{\Delta}}\right)\right]_{x_-}^{x_+}. \tag{3.108}$$

The values of $x_\pm$ are solutions of $Ax^2 + Bx + C = 0$, that are

$$x_\pm = \frac{-B \mp \sqrt{\Delta}}{2A}, \tag{3.109}$$

so that $r_- = \sqrt{x_-}$, $r_+ = \sqrt{x_+}$. Substituting in Eq. 3.108 x_- and x_+ as from (3.109), we get

$$T_r = -\frac{\sqrt{2}}{2\sqrt{-A}}[\arcsin(-1) - \arcsin(1)] = \frac{\sqrt{2}\pi}{2\sqrt{-A}} = \frac{\pi}{2}\frac{1}{\sqrt{\frac{\pi}{3}G\rho_0}}, \tag{3.110}$$

which is the same results of the case $L^2 = 0$. This is relevant because it means that T_r does not depend neither on E nor on L^2, so it corresponds to a fully *isochrone* radial oscillation. Actually, the oscillation period depends only upon ρ_0, other than (as obvious) upon G. This characteristic feature reminds that of the small oscillations of a simple pendulum whose period does not depend on the initial conditions (and so on the energy and angular momentum) but just on the length of the wire and on the gravitational acceleration.

Exercise 3.6 In the case of a homogeneous sphere, calculate T_r for orbits such that $r_- < R \le r_+$ (case (ii) of (3.101)).

Solution Left to the reader.

3.6.3 Radial Period in the Plummer Potential

A Plummer sphere of mass M and scale radius b is characterized by the potential (see Eq. 3.50)

$$U(r) = \frac{GM}{\sqrt{r^2 + b^2}}, \tag{3.111}$$

corresponding to the effective potential energy

$$V_{eff}(r; L) = -\frac{GM}{\sqrt{r^2 + b^2}} + \frac{1}{2}\frac{L^2}{r^2} \tag{3.112}$$

Consequently, the radial period of oscillations for given E and L is

$$T_r = \sqrt{2} \int_{r_-}^{r_+} \frac{dr}{\sqrt{E + \frac{GM}{\sqrt{r^2+b^2}} - \frac{1}{2}\frac{L^2}{r^2}}}. \tag{3.113}$$

Note that solving the non-linear equation $E - V_{eff}(r; L^2) = 0$ to determine r_- and r_+ requires a numerical approach.

The integral in Eq. 3.113, after proper change of variable, can be, in the general case $L^2 > 0$, put in the form of a combination of *elliptic integrals* of the 1st and 3rd kind. If $L = 0$ (radial orbits, $r_- = 0$) the integral is simpler and can be expressed as an elliptical integral of the 1st kind. The study of these special functions is out of the scopes of this book, for which I address interested readers to the classical book [5]. In any case, obtaining T_r requires a numerical treatment and often it is more convenient using methods of numerical quadrature.

Given this, we limit here to the study of the case of small amplitude oscillations of radial orbits ($L^2 = 0$) around the sphere centre, i.e. the case of low energies, $E \gtrsim -GM/b$. For $L^2 = 0$, and for any E, we have $r_- = 0$ and $r_+ = b\sqrt{\left(\frac{GM}{bE}\right)^2 - 1}$, and

$$T_r = \sqrt{2} \int_{0}^{r_+} \frac{dr}{\sqrt{E + \frac{GM}{\sqrt{r^2+b^2}}}}, \tag{3.114}$$

which, by a Taylor expansion of the integrand up to 4th order, gives

$$T_r = \frac{\sqrt{2}}{\sqrt{E + (GM)/b}} \left[Ar_+ + \frac{B}{3}r_+^3 - \frac{C}{5}r_+^5 + O(r_+^7) \right], \tag{3.115}$$

where

$$\begin{cases} A = 1, \\ B = \dfrac{GM}{4b^2\,(Eb+GM)}, \\ C = \dfrac{3GM\,(2Eb+GM)}{32b^4\,(Eb+GM)^2}. \end{cases} \tag{3.116}$$

Finally, the radial period assumes the expression

$$T_r = \frac{\sqrt{2}}{\sqrt{E+(GM)/b}}r_+ \times \left[1+\frac{GM}{12b^2(Eb+GM)}r_+^2 - \frac{3GM(2Eb+GM)}{160b^4(Eb+GM)^2}r_+^4 + O(r_+^6)\right], \tag{3.117}$$

which, at lowest order, approximates as

$$T_r = \frac{\sqrt{2}}{\sqrt{E+(GM)/b}}r_+ = \frac{\sqrt{\left(\frac{GM}{bE}\right)^2-1}}{\sqrt{E+(GM)/b}}\sqrt{2}b. \tag{3.118}$$

3.6.4 *Radial Period in the Isochrone Potential*

As we saw in Sect. 3.4.2, the isochrone potential summarizes some characteristics of a uniform density matter distribution (at small radii) and of a Keplerian potential at large radii. Actually, the isochrone potential is (Eq. 3.56)

$$U(r) = \frac{GM}{b+\sqrt{b^2+r^2}} = \frac{GM}{b}\frac{1}{1+\sqrt{1+(r/b)^2}}, \tag{3.119}$$

where the length scale, b, and the total mass, M, are the 2 free parameters of the model.

The radial period of bound oscillations (that requires $V_{eff,min} < E < 0$) is

$$T_r = \sqrt{2}\int\limits_{r_-}^{r_+} \frac{\mathrm{d}r}{\sqrt{E+\frac{GM}{b}\frac{1}{1+\sqrt{1+(r/b)^2}}-\frac{1}{2}\frac{L^2}{r^2}}}. \tag{3.120}$$

With the change of variables

$$s = \frac{GM}{b\,U(r)} = 1 + \sqrt{1 + \left(\frac{r}{b}\right)^2} \geq 2, \tag{3.121}$$

it results

$$\mathrm{d}r = b\frac{s-1}{\sqrt{s(s-2)}}\mathrm{d}s, \tag{3.122}$$

and the integral in Eq. 3.120 transforms into

$$T_r = \sqrt{2}b \int_{s_-}^{s_+} \frac{(s-1)\mathrm{d}s}{\sqrt{Es^2 - \left(2E - \frac{GM}{b}\right)s - 2\frac{GM}{b} - \frac{1}{2}\frac{L^2}{b^2}}}, \tag{3.123}$$

that, indicating with s_- and s_+ ($s_- < s_+$) the two roots of the denominator, can be rewritten as

$$T_r = b\sqrt{\frac{2}{-E}} \int_{s_-}^{s_+} \frac{(s-1)}{\sqrt{(s-s_-)(s_+-s)}} ds = \pi b\sqrt{\frac{2}{-E}}\left[-1 + \frac{1}{2}(s_- + s_+)\right]. \tag{3.124}$$

As it is easily seen:

$$\frac{1}{2}(s_- + s_+) = 1 - \frac{GM}{2bE}, \tag{3.125}$$

and finally

$$T_r = \frac{\sqrt{2}}{2}\pi\frac{GM}{(-E)^{3/2}}. \tag{3.126}$$

Note that this result (independent of both b and L^2, justifying the adjective *isochronous*) is the same as for the Keplerian potential of a mass M (Eq. 3.100).

3.7 Azimuthal Period

As for the radial period, we can define an *azimuthal* period as the (constant) time needed to make one full revolution around the centre of force, i.e. the time needed to increase of 2π radians the angular coordinate, θ, on the motion plane.

Formally, T_θ is given by

$$T_\theta = \int_{t(\theta)}^{t(\theta+2\pi)} \mathrm{d}t = \int_{\theta}^{\theta+2\pi} \frac{\mathrm{d}\theta}{\dot{\theta}} = \frac{1}{L}\int_{\theta}^{\theta+2\pi} r^2\mathrm{d}\theta = \int_{r(\theta)}^{r(\theta+2\pi)} \frac{\mathrm{d}r}{\dot{r}}, \tag{3.127}$$

where we made use of the angular momentum conservation, $L = r^2\dot{\theta} = r^2\frac{\mathrm{d}\theta}{\mathrm{d}r}\dot{r} = const$. Recall that for the choice of coordinates done throughout this book (see Sect. 3.2), it is $L_z = L \geq 0$. To evaluate T_θ, we can choose as initial θ the pericentre azimuth, for which $r(\theta) \equiv r_-$. Of course, the knowledge of the polar expression for the trajectory, $r(\theta)$, would allow—apart possible complications in the computation—to obtaining the azimuthal period as straight evaluation of the integral

$$T_\theta = \frac{1}{L}\int\limits_{\theta}^{\theta+2\pi} r^2(\theta)\mathrm{d}\theta. \tag{3.128}$$

This is often unviable because of the difficulty in finding an explicit expression for $r(\theta)$. So, we examine a better way to get T_θ by the knowledge of the potential of the force field, for given values of E and L.

To accomplish the task, we, first, define as $\Delta\theta$ the azimuthal angular variation of the particle in a complete radial oscillation, lasting a time T_r. This corresponds to write

$$\Delta\theta = \int\limits_{\theta(r_-)}^{\theta(r_+)} \mathrm{d}\theta + \int\limits_{\theta(r_+)}^{\theta(r_-)} \mathrm{d}\theta = 2\int\limits_{r_-}^{r_+} \frac{L}{r^2}\frac{\mathrm{d}r}{\sqrt{2[E - V_{eff}(r; L)]}}. \tag{3.129}$$

From its definition, it is clear that:

$$\begin{cases} \Delta\theta \geq 2\pi, & \text{if } T_\theta \leq T_r, \\ \Delta\theta \leq 2\pi, & \text{if } T_\theta \geq T_r. \end{cases} \tag{3.130}$$

The case $\Delta\theta = 2\pi$ corresponds to a closed orbit (like Keplerian); the cases $\Delta\theta > 2\pi$ or $\Delta\theta < 2\pi$ to an orbital precession, either prograde (forward) or retrograde (backward).

Note that the integrand in Eq. 3.129 above is $L/r^2 = \dot{\theta}$ times the integrand involved in the expression of T_r so that, by the integral mean theorem, $\langle\dot{\theta}\rangle = \Delta\theta/T_r = const.$, where $\langle\dot{\theta}\rangle$ is the average angular velocity along the radial oscillation, i.e.

$$\langle\dot{\theta}\rangle = \frac{1}{T_r}\int\limits_0^{T_r} \dot{\theta}\mathrm{d}t = const. \tag{3.131}$$

This latter relation, valid because $\dot{\theta} > 0,\ \forall t$ and the motion is radially periodic, leads to

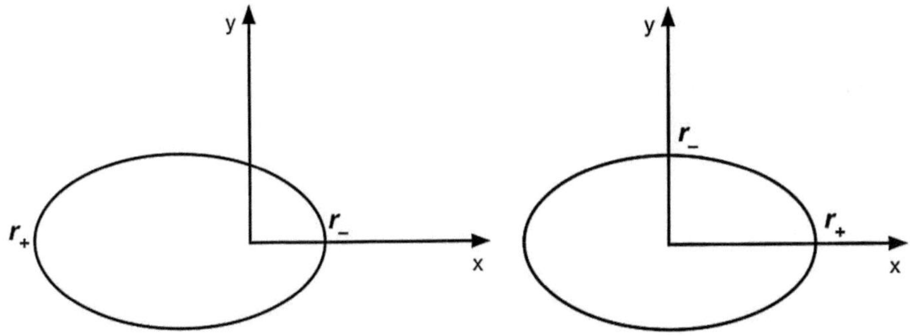

Fig. 3.6 A Keplerian ellipse (left panel; the origin is at one of the foci) and an elliptic trajectory in a homogeneous sphere (right panel; the origin is at the centre of the homogeneous sphere)

$$T_\theta = \frac{2\pi}{\langle\dot{\theta}\rangle} = \frac{2\pi}{\Delta\theta} T_r. \tag{3.132}$$

For the Keplerian and the homogeneous sphere cases, $T_r = T_\theta$ and $T_r = T_\theta/2$, respectively (Fig. 3.6).

3.7.1 Fully Periodic Motion

From what we said above, a planar motion in an external radial force field may be *radially* periodic and/or *azimuthally* periodic, and this happens when there exists a constant time, interval T_r such that $r(t + T_r) = r(t)$ and another constant time interval T_θ such that $\theta(t + T_\theta) = \theta(t)$, for every t.

A *full* (or simple) periodicity corresponds to the existence of a single period T such that $\mathbf{r}(t + T) = \mathbf{r}(t)$ for all t. For planar motion, this condition is equivalent to $r(t + T) = r(t)$ and $\theta(t + T) = \theta(t)$.

It is easy to see that if T_r and T_θ are *commensurable*, i.e. their ratio is a rational number, $T_\theta/T_r = l/m$ where l and m are positive integers, then the motion is fully periodic and the period is $T = lT_r = mT_\theta$.

Let us show, indeed, that, upon the above hypotheses, it results

$$r(t + T) = r(t), \text{ and } \theta(t + T) = \theta(t), \ \forall t. \tag{3.133}$$

Actually, if T_r is the radial period, i.e. $r(t+T_r) = r(t)$, then $r(t+T) = r(t+lT_r) = r(t+(l-1)T_r+T_r) = r(t+(l-1)T_r) = r(t+(l-2)T_r+T_r) = r(t+(l-2)T_r) = r.\cdots = r(t + (l - l + 1)T_r + T_r) = r(t + T_r) = r(t)$,

By the same token, it can be shown that $\theta(t + T) = \theta(t + mT_\theta) = \theta(t)$ and so the periodicity is proven.

3.8 The Inverse Problem in a Central Force Field

In a central force field the equiforce surfaces are spherical, $\mathbf{F} = f(r)\,\mathbf{e}_r$, the motion is planar, and the equations of motions in the plane of motion are

$$\begin{cases} \mathbf{a}_r &= \left(\ddot{r} - r\dot{\theta}^2\right)\mathbf{e}_r = f(r)\,\mathbf{e}_r, \\ \mathbf{a}_\theta &= \left(r\ddot{\theta} + 2\dot{r}\dot{\theta}\right)\mathbf{e}_\theta = 0. \end{cases} \tag{3.134}$$

As we saw in Sect. 3.2, integration of the second of the equations above leads to $L_z = r^2\dot{\theta} = const.$, so that we can eliminate $\dot{\theta}$ in the first equation.
Actually, letting again $L = L_z$,

$$\dot{r} = \frac{\mathrm{d}r}{\mathrm{d}\theta}\dot{\theta} = \frac{\mathrm{d}r}{\mathrm{d}\theta}\frac{L}{r^2}, \tag{3.135}$$

and

$$\ddot{\theta} = \frac{\mathrm{d}\dot{\theta}}{\mathrm{d}t} = \frac{L}{r^2}\frac{\mathrm{d}}{\mathrm{d}\theta}\left(\frac{L}{r^2}\right) = -2\frac{L^2}{r^5}\frac{\mathrm{d}r}{\mathrm{d}\theta}, \tag{3.136}$$

implying

$$\ddot{r} = \frac{\mathrm{d}}{\mathrm{d}t}\left(\frac{\mathrm{d}r}{\mathrm{d}\theta}\dot{\theta}\right) = \frac{\mathrm{d}^2 r}{\mathrm{d}\theta^2}\dot{\theta}^2 + \frac{\mathrm{d}r}{\mathrm{d}\theta}\ddot{\theta} = \left(\frac{L}{r^2}\right)^2\frac{\mathrm{d}^2 r}{\mathrm{d}\theta^2} - 2\frac{L^2}{r^5}\left(\frac{\mathrm{d}r}{\mathrm{d}\theta}\right)^2. \tag{3.137}$$

Excluding radial orbits ($L = 0$), which can cross the origin, and letting $u \equiv \dfrac{1}{r}$, we have

$$\frac{\mathrm{d}r}{\mathrm{d}\theta} = -\frac{1}{u^2}\frac{\mathrm{d}u}{\mathrm{d}\theta} \longrightarrow \frac{\mathrm{d}^2 r}{\mathrm{d}\theta^2} = \frac{2}{u^3}\left(\frac{\mathrm{d}u}{\mathrm{d}\theta}\right)^2 - \frac{1}{u^2}\frac{\mathrm{d}^2 u}{\mathrm{d}\theta^2}, \tag{3.138}$$

which leads to writing the first of Eqs. 3.134 in terms of u and θ

$$\begin{aligned} L^2 u^4\left[\frac{2}{u^3}\left(\frac{\mathrm{d}u}{\mathrm{d}\theta}\right)^2 - \frac{1}{u^2}\frac{\mathrm{d}^2 u}{\mathrm{d}\theta^2}\right] - 2L^2 u\left(\frac{\mathrm{d}u}{\mathrm{d}\theta}\right)^2 - L^2 u^3 = \\ = -L^2\left[u^2\frac{d^2 u}{d\theta^2} + u^3\right] = f(1/u). \end{aligned} \tag{3.139}$$

The final form of the equation for the trajectory is so

$$u^2\left(\frac{\mathrm{d}^2 u}{\mathrm{d}\theta^2} + u\right) = -\frac{1}{L^2}f(1/u). \tag{3.140}$$

Given the force field $f(1/u)$, the solution of the above non-linear differential equation for the unknown function $u(\theta)$ would give the trajectory $r(\theta) = 1/u(\theta)$ of the body, once that $u(\theta_0)$ and $\left(\frac{\mathrm{d}u}{\mathrm{d}\theta}\right)_0$ are provided. Both these conditions are obtained from initial (in time) conditions $r(0) = r_0$ and $\dot{r}(0) = \dot{r}_0$, being (with the choice θ_0 as the anomaly at time $t = 0$)

$$u(\theta_0) = \frac{1}{r_0}, \qquad \text{and} \qquad \left(\frac{\mathrm{d}u}{\mathrm{d}\theta}\right)_0 = -\frac{1}{r_0^2}\left(\frac{\mathrm{d}r}{\mathrm{d}\theta}\right)_0 = -\frac{\dot{r}_0}{L}. \tag{3.141}$$

The search of solution for $r(\theta)$ once given $f(r)$ corresponds, indeed, to the *direct* problem of dynamics in a central force field. On the other side, assuming known the planar trajectory $r(\theta)$, the search for a central force field compatible with this trajectory constitutes the *inverse* problem of dynamics.

The existence and uniqueness of a solution for this inverse problem is not guaranteed. Under some conditions the central force field compatible with a given family of trajectories is given by

$$f(1/u) = -L^2u^2\left(\frac{d^2u}{d\theta^2} + u\right). \tag{3.142}$$

Of course, this explicit solution is found whenever the trajectory $u(\theta)$ is twice differentiable and the right hand side of Eq. 3.142 can be expressed as explicit function of u (which is not always the case).

3.8.1 From Elliptic Trajectories to their Central Force Field

A Keplerian ellipse is an elliptical trajectory one focus of which is in the coordinate origin (where the massive point is located). Its polar expression is therefore given by

$$r(\theta) = \frac{p}{1 + e\cos(\theta - \theta_0)} = \frac{1}{u}, \tag{3.143}$$

where $e < 1$ is the eccentricity, p the ellipse semi-latus rectum and θ_0 the angular position of the pericentre (indeed, $r_- = r(\theta_0) = p/(1+e)$, while $r_+ = r(\theta_0+\pi) = p/(1-e)$). The choice $\theta_0 = 0$ is commonly made, which means counting θ angles starting from the pericentre position.

For the given elliptic trajectory, it is

$$\frac{\mathrm{d}u}{\mathrm{d}\theta} = -\frac{e\sin(\theta - \theta_0)}{p} \qquad \Longrightarrow \qquad \frac{\mathrm{d}^2u}{\mathrm{d}\theta^2} = -\frac{e\cos(\theta - \theta_0)}{p}, \tag{3.144}$$

so that Eq. 3.143 gives

$$f(1/u) = -L^2u^2\left(-\frac{e\cos(\theta-\theta_0)}{p} + \frac{1+e\cos(\theta-\theta_0)}{p}\right) = -\frac{L^2}{p}u^2, \tag{3.145}$$

which, in terms of r, gives the Keplerian force field

$$f(r) = -\frac{L^2}{p}\frac{1}{r^2}. \tag{3.146}$$

On the other hand, ellipses centred in the origin have the so-called "canonical" (Cartesian) form

$$\frac{x^2}{a^2} + \frac{y^2}{b^2} = 1. \tag{3.147}$$

In planar polar coordinates, the equation above takes the form

$$\begin{cases} x = a\cos\theta, \\ y = b\sin\theta. \end{cases} \tag{3.148}$$

Thanks to (3.148), and letting $u \equiv 1/r$, Eq. 3.147 transforms into

$$u^2 = \frac{1}{a^2\cos^2\theta + b^2\sin^2\theta}. \tag{3.149}$$

Differentiating the above expression, we get

$$\frac{\mathrm{d}u^2}{\mathrm{d}\theta} = 2u\frac{\mathrm{d}u}{\mathrm{d}\theta} = 2(a^2-b^2)\sin\theta\cos\theta u^4, \tag{3.150}$$

which leads, in sequence, to

$$\frac{\mathrm{d}u}{\mathrm{d}\theta} = (a^2-b^2)\sin\theta\cos\theta u^3, \tag{3.151}$$

and to

$$\frac{\mathrm{d}^2u}{\mathrm{d}\theta^2} = (a^2-b^2)\left[3(a^2-b^2)u^5\sin^2\theta\cos^2\theta + u^3(\cos^2\theta - \sin^2\theta)\right]. \tag{3.152}$$

Given the above expression, Eq. 3.142 becomes

$$f(1/u) = -L^2u^2\Big\{(a^2-b^2)\Big[3u^5(a^2-b^2)\sin^2\theta\cos^2\theta+ \\ +u^3(\cos^2\theta-\sin^2\theta)\Big]+u\Big\}, \tag{3.153}$$

and so, eliminating $\sin^2\theta$ and $\cos^2\theta$ by means of Eq. 3.149, we finally have

$$f(1/u) = -\frac{L^2}{a^2b^2}\frac{1}{u} = -\frac{L^2}{a^2b^2}r. \tag{3.154}$$

We have thus obtained the harmonic force, $f(r) \propto -r$, as expected.

Exercise 3.7 In spherical symmetry, what potential admits trajectories of type $r = ae^{b\theta}$, with $a > 0$ and $b \in \mathbb{R}$ (*logarithmic spiral*)?

Solution

$$f\left(\frac{1}{u}\right) = -L^2\frac{1}{r^2}\left(\frac{b^2}{a}\mathrm{e}^{-b\theta}+\frac{1}{r}\right) = -L^2\frac{1}{r^2}\left(\frac{b^2}{r}+\frac{1}{r}\right) = \\ = -L^2\frac{1}{r^3}\left(b^2+1\right) = \frac{dU}{dr}.$$

By straightforward integration

$$U(r) = \frac{(b^2+1)L^2}{2}\frac{1}{r^2}+c,$$

from which, using Poisson's equation, we get

$$\rho(r) = \frac{1}{4\pi G}\frac{1}{r^2}\frac{\mathrm{d}}{\mathrm{d}r}\left[r^2L^2\left(b^2+1\right)\frac{1}{r^3}\right] = \\ = \frac{1}{4\pi G}L^2\left(b^2+1\right)\frac{1}{r^2}\left(-\frac{1}{r^2}\right) < 0, \qquad \forall b \in \mathbb{R}.$$

The above density is always negative, that means that the logarithmic spiral trajectories are not compatible with a dissipationless Newtonian gravitational field.

3.9 Further Readings

The contents of this chapter are strictly related to the ones of the previous chapter. So the relevant additional readings partly overlap, and the suggested books are [1, 3, 4, 6, 7, 9–14].

References

1. Arnold, V.I.: Mathematical Methods of Classical Mechanics, 2nd edn. Springer, New York (1989)
2. Bertrand, J.: Théoréme relatif au mouvement d'un point attiré vers un centre fixe. C. R. Acad. Sci. **77**, 849–853 (1873)
3. Binney, J., Tremaine, S.: Galactic Dynamics, 2nd edn. Princeton University Press, Princeton (2011)
4. Boccaletti, D., Pucacco, G.: Theory of Orbits, vols. 1 and 2. Springer, Berlin/Heidelberg (1996)
5. Byrd, P.F., Friedman, M.D.: Handbook of Elliptic Integrals for Engineers and Scientists, 2nd edn. revised. Springer, Berlin/Heidelberg/New York (1971)
6. Fridman, A.M., Polyachenko, V.L.: Physics of Gravitating Systems I and II. Springer, Berlin (2012)
7. Goldstein, H., Safko, J.L., Poole, G.P., Spiegel, M.R.: Classical Mechanics, 3rd edn. Pearson Ed. Limited, Edinburgh (2002)
8. Hénon, M.: L'amas isochrone. Annales d'Astrophysique **22**, 126 (1959)
9. Landau, L.D., Lifshits, E.M.: Mechanics, Course of Theoretical Physics, vol. 1, 3rd edn. Elsevier Butterworth-Heinemann, Oxford (1976)
10. Landau, L.D., Lifshits, E.M.: The Classical Theory of Fields, Course of Theoretical Physics, vol. 2, 4th edn. Elsevier Butterworth-Heinemann, Oxford (1987)
11. Mc Call, M.W.: Classical Mechanics: From Newton to Einstein: A Modern Introduction, 2nd edn. J.Wiley & Sons Ltd., Chichester (2011)
12. Merritt, D.: Dynamics and Evolution of Galactic Nuclei. Princeton University Press, Princeton (2013)
13. Misner, C.W., Thorne, K.S., Wheeler, J.A.: Gravitation. W. H. Freeman and Company, New York (1973)
14. Saslaw, W.C.: Gravitational Physics of Stellar and Galactic Systems. Cambridge University Press, Cambridge (1985)

Chapter 4
Astrophysical Applications

This chapter is dedicated to applications of classical Newtonian gravity in the astronomical and astrophysical context. The main difference with respect to the role of gravity on earth is that, in the cosmos, the gravitational field is not constant, either in magnitude or in direction. This distinction is crucial and is rooted in the concept of self-gravity. We analyse both equilibrium and non-equilibrium properties of self-gravitating systems under the simplifying assumption of spherical symmetry.

4.1 Terrestrial and Astrophysical Contexts

The main difference between fluids in terrestrial and astrophysical context is given by (i) the range of variation of density, pressure and temperature involved and (ii) the role played by the body forces.

Regarding (i), on earth the densities, pressures and temperatures are limited to a relatively narrow range around the characteristic environment values although, for example, air density reduces a lot at high altitudes, by a factor of ten when going from ground up to 17 km of altitude, as well as the temperature of lava in volcanoes exceeds 1000 °C, compared to the $\sim$20 °C of environmental air. As another example, the pressure in water depends of course on the depth of water where we measure it. In the deepest regions of oceans (depth around 11,000 metres under the sea surface) pressure reaches 1100 kg cm^{-2} and temperature drops to just above 0 °C.

On the other side, in astrophysical fluids the range of variation of the state variables is very huge, densities going from almost nuclear values, $\sim 10^{14}$ g cm^{-3} in the centre of collapsed dead stars like neutron stars to values below 10^{-21} g cm^{-3} (i.e. $\sim 10^3$ particles per cubic centimetre) in the Large Molecular Clouds (LMCs), sites of star formation. Temperatures varies from few Kelvins in the LMC to the

R. A. Capuzzo Dolcetta, *Classical Newtonian Gravity and Potential Theory*, UNITEXT for Physics, https://doi.org/10.1007/978-3-032-13036-5_4

$\sim 10^8$ K in the nuclei of helium burning stars. Analogously, pressures span more than eight orders of magnitude.

The point (ii) is that of interest for this book and regards essentially the gravitational, long range, body force. This force is present everywhere in the physical world but its role is quite different when dealing with the study of dynamics of objects on earth with respect to what happens in an astrophysical environment. This chapter deals essentially with the distinction between terrestrial systems, which gravitate in an external fixed gravitational field, and astrophysical objects, which gravitate in their own self-generated gravitational field.

4.2 Self-Gravitating Systems

The concept of *self-gravitation* forms the basis of this chapter. On earth, flows are essentially determined by the combined roles of local (small scale) forces and the large scale ones, due to gravity and Coulomb forces. On earth, gravity is for all practical purposes a body force of constant radial direction and intensity. We can say that liquids like lakes and rivers are gravitating in an external *fixed* field. The flow of a river is caused by its (approximate) free fall (out of equilibrium) from the source to the mouth, while lakes have just local movements around an equilibrium between the terrestrial gravity and the binding reaction of the ground. Astrophysical fluids, instead, are such that the body forces may vary significantly within their volume. Even considering gravity, only, it is different from one point to another. Astrophysical fluids are essentially self-gravitating fluids while terrestrial are not. We can clarify this better.

As we know, in a fluid, and in general in any system of N particles, all the constituents exert forces on each other, and the distinction between short range forces and long range forces leads in fluid dynamics, to the distinction between surface and body forces. Newtonian gravity is long range and so every ith particle interacts with any other jth via a force scaling as r_{ij}^{-2} where r_{ij} is the distance between the objects in the pair. Thus, the generic ith particle is subjected to an *internal* gravity acceleration due to the presence/interaction of/with the other $N-1$ particles added to an *external* gravity acceleration due to the presence of external massive bodies. In formulas

$$\ddot{\mathbf{r}}_i = G \sum_{\substack{j=1 \\ j \neq i}}^{N} \frac{m_j}{r_{ij}^3} (\mathbf{r}_j - \mathbf{r}_i) + \mathbf{F}_i^{(ext)}. \tag{4.1}$$

Clearly, if for every particle i the acceleration induced by the other particles of the system is large with respect to the external acceleration the fluid is said *self-gravitating*, and vice versa. The self-gravity condition can be tested by checking an energetic condition. The energetic condition consists in an approximated evaluation of the dimensionless parameter $\alpha = |\Omega_{int}|/|\Omega_{ext}|$ where Ω_{int} and Ω_{ext} stands, respectively, for the values of gravitational internal and external energy. The limit $\alpha \ll 1$ would be the one of non-self-gravitating fluids.

The precise evaluation of Ω_{int} is practically impossible because it involves a double summation over all the particles of the fluids

$$\Omega_{int} = -\frac{1}{2} \sum_{\substack{(i,j)=1 \\ i \neq j}}^{N} G \frac{m_i m_j}{r_{ij}}, \tag{4.2}$$

so for practical use it is necessary to resort to an approximate evaluation, which we give here as an example.

Consider a lake like, for instance, the big lake Victoria crossing Kenya, Uganda and Tanzania in Africa. Its basin has a volume of 2760 km^3 $= 2.76 \times 10^{18}$ cm^3. This means a mass of water of $M = 2.76 \times 10^{18}$ g. A rough, but valid, evaluation of the energy of self-gravitation of the lake can be obtained by calculating the gravitational energy of the same mass of water enclosed in a sphere whose radius is, obviously, $R = [3M/(4\pi\rho)]^{1/3} \simeq 8.7018$ km, where $\rho = 1$ g cm^{-3} is the density of water. A straightforward calculation of the integral giving the gravitational energy of a uniform sphere of mass M and radius R gives

$$\Omega_{int} = -\frac{3}{5}\frac{GM^2}{R} \simeq -3.50 \times 10^{23}\,\text{erg} = -3.50 \times 10^{16}\,\text{J}, \tag{4.3}$$

while the external (Earth) gravitational energy is

$$\Omega_{ext} = -\frac{GMM_E}{R_E} \simeq -1.73 \times 10^{30}\,\text{erg} = -1.73 \times 10^{23}\,\text{J}, \tag{4.4}$$

where M_E and R_E are the mass and equatorial radius of the Earth. Consequently,

$$\alpha = \frac{3}{5}\frac{M}{M_E}\frac{R_E}{R}, \tag{4.5}$$

which, assuming $M_E = 5.972 \times 10^{24}$ kg and $R_E = 6371$ km gives $\alpha \simeq 2.0302 \times 10^{-7}$ which is $\ll 1$. So, fluids on earth are actually *non self-gravitating*. On the contrary, astrophysical objects like stars are self-gravitating because they are all composed of fluid (essentially plasma, in a star like the Sun) which is "contained" by its own internal gravity while the external gravity is negligible. The real difference is, indeed, that in stars there is not a solid/rigid container of a fluid that, for terrestrial gases, is necessary to confine the gas against its pressure gradient, but, instead, this confinement role is played by the strong self-gravitational force.

Self-gravity represents a relevant complication in the study of fluids because in the governing equations it gives a body force field which is not a constant vector but depends on the position and, in non-stationary cases, on time also. This vector field, to be determined, needs the solution of the proper field equation which links the source of field (the matter density ρ) with the potential it generates, $U(\mathbf{r})$. In the case of gravity, this is the Poisson's equation

$$\nabla^2 U = -4\pi G\rho, \tag{4.6}$$

which must be added to the whole system of constitutive equations. Consequently, the full system of constitutive equations for variables ρ, pressure, p, internal energy per unit mass, e, potential U, and velocity, $\mathbf{v}$, for an astrophysical fluid—when neglecting dynamical electromagnetic effects—can be written, in Eulerian form, as

$$\begin{cases} \dfrac{\partial \rho}{\partial t} + \nabla \cdot \rho \mathbf{v} = 0 \\[2ex] \dfrac{\partial \mathbf{v}}{\partial t} + (\mathbf{v} \cdot \nabla)\, \mathbf{v} = -\dfrac{1}{\rho} \nabla p + \dfrac{\eta}{\rho} \nabla^2 \mathbf{v} + \left(\dfrac{1}{3} \dfrac{\eta}{\rho} + \dfrac{\zeta}{\rho} \right) \nabla \nabla \cdot \mathbf{v} + \nabla U \\[2ex] \dfrac{\partial}{\partial t} \rho \left(\dfrac{1}{2} v^2 + e \right) + \nabla \cdot \rho \left(\dfrac{1}{2} v^2 + e + \dfrac{p}{\rho} \right) \mathbf{v} = \\ \qquad = \left[\dfrac{\eta}{\rho} \nabla^2 \mathbf{v} + \left(\dfrac{1}{3} \dfrac{\eta}{\rho} + \dfrac{\zeta}{\rho} \right) \nabla \nabla \cdot \mathbf{v} \right] \cdot \mathbf{v} + \rho \dfrac{\mathrm{D} q^*}{\mathrm{D} t} \\[2ex] \nabla^2 U = -4\pi G \rho \\[2ex] f(\rho, p, e) = 0, \end{cases} \tag{4.7}$$

where q^* in the 3rd equation (the *energy* equation) is the quantity of heat per unit mass. The 1st and 2nd equation in the system are, respectively, the *continuity* (or mass conservation) and Navier-Stokes' equation of motion, in presence of shear (quantified by the η coefficient) and bulk (ζ coefficient) viscosity, and the last equation is a functional relation among state variables, called equation of state (EOS).

Equations 4.7 are six partial differential equations and an algebraic equation for seven unknowns (the four scalar fields ρ, p, e, U, and the vector $\mathbf{v}$ field) subjected to proper initial and boundary conditions.

The presence of self-gravity is not the only difference in the treatment of astrophysical fluids and flows with respect to earth. Examples of non self-gravitating and self-gravitating systems are shown in Figs. 4.1 and 4.2.

Another relevant difference is the energy generation and transport in the fluid.

In stars, energy is generated by nuclear reactions in their innermost region which is characterized by high pressure and temperature. This energy is released slowly in time but in huge and long lasting quantity. The energy output of nuclear reactions is in the form of radiation and kinetic energy of resulting particles and neutrinos. The energy transport, crucially depending on the interaction with the surrounding

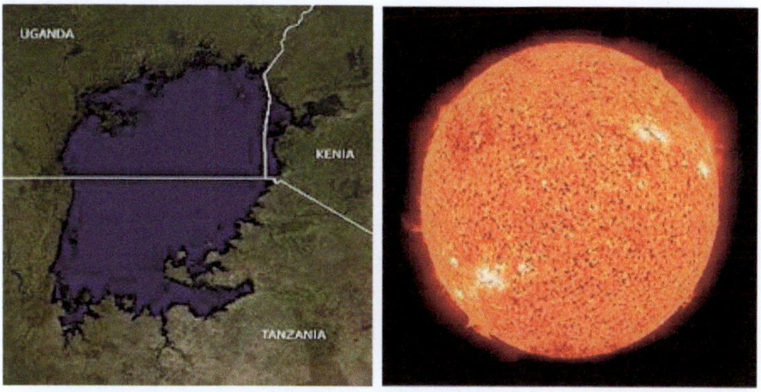

Fig. 4.1 A typical non self-gravitating fluid (lake Victoria, on the left) and a typical self-gravitating one (the Sun, on the right)

Fig. 4.2 The M13 globular cluster (left) and a bright galaxy cluster (right)

matter, determines the internal structure of the star. So, both in stars and in other astrophysical environments, the energy transport should be properly accounted for in the energy equation of the fluid, that is in the 3rd equation in the system of Eqs. 4.7. As a matter of fact, while the fundamental equations of fluid dynamics are in common in the earth and astrophysical (non-relativistic) environments, in the astrophysical context fluid-dynamics involves a series of specifities that characterize what is commonly known as Astrophysical fluid dynamics. We give below important examples of such peculiarities involving Newtonian gravity.

4.2.1 *Equilibrium of Self-Gravitating Systems*

The inclusion of Poisson's equation as coupled to Euler's equation of motion, together with the assumption of spherical symmetry for density and pressure, leads to

$$\begin{cases} \dfrac{\mathrm{D}v_r}{\mathrm{D}t} = -\dfrac{1}{\rho}\dfrac{\mathrm{d}p}{\mathrm{d}r} + \dfrac{\mathrm{d}U}{\mathrm{d}r} \\ \\ \dfrac{\mathrm{d}U}{\mathrm{d}r} = -G\dfrac{M(r)}{r^2} \end{cases} \tag{4.8}$$

where v_r is the radial component of the velocity field (here $v_\theta = v_\phi = 0$), $\mathrm{D}v_r/\mathrm{D}t$ is its Lagrangian derivative (derivative along the flow), and $M(r)$ is the mass enclosed in the sphere of radius r, namely

$$M(r) = 4\pi \int_0^r \rho r^2 \,\mathrm{d}r. \tag{4.9}$$

At equilibrium, $\mathrm{D}v_r/\mathrm{D}t = 0$ and, accounting for Eq. 4.9, the two equations in (4.8) reduce to the single integro-differential equation

$$\frac{1}{\rho}\frac{\mathrm{d}p}{\mathrm{d}r} = -G\frac{M(r)}{r^2}. \tag{4.10}$$

This equation can be transformed into a 2nd order differential equation by multiplication of both sides by r^2, followed by a differentiation respect to r, to obtain

$$\frac{\mathrm{d}}{\mathrm{d}r}\left(\frac{r^2}{\rho}\frac{\mathrm{d}p}{\mathrm{d}r}\right) = -4\pi G\rho r^2, \tag{4.11}$$

which is still in two unknowns (ρ and p). As usual, the reduction to just one unknown is done by using an EOS, provided it is , i.e. in the form $p = p(\rho)$.

Under such assumption, Eq. 4.11 becomes

$$\frac{\mathrm{d}}{\mathrm{d}r}\left(\frac{r^2}{\rho}\frac{\mathrm{d}p}{\mathrm{d}\rho}\frac{\mathrm{d}\rho}{\mathrm{d}r}\right) = -4\pi G\rho r^2. \tag{4.12}$$

Let us examine two particular cases of EOS: the isothermal case ($T = const. = T_0$) and the polytropic case $p \propto \rho^\gamma$, which indeed contains, for $\gamma = 1$, the isothermal case.

The EOS of an astrophysical gas, whenever pressure is not too high and for sufficiently high temperature, can be acceptably represented with its ideal gas expression $p = nkT = \rho kT/(\mu m_H)$.

In the (spatial) isothermality assumption the ideal gas EOS becomes linear in ρ because μ shows a much weaker dependence on ρ and on the specific abundances of various chemical species than on T, so that μ can be considered constant once fixed T_0, and the equilibrium equation 4.12 becomes

$$\frac{kT_0}{\mu_0 m_H} \frac{\mathrm{d}}{\mathrm{d}r} \left(\frac{r^2}{\rho} \frac{\mathrm{d}\rho}{\mathrm{d}r} \right) = -4\pi G \rho r^2, \tag{4.13}$$

that can be also expressed in logarithmic form as

$$\frac{kT_0}{\mu_0 m_H} \frac{\mathrm{d}}{\mathrm{d}r} \left(r \frac{\mathrm{d}\ln\rho}{\mathrm{d}\ln r} \right) = -4\pi G \rho r^2. \tag{4.14}$$

An explicit solution of the above equation can be found in power law form, $\rho = a^2 r^\alpha$, where a and α have to be determined (to have equilibrium α must be negative).[1] Substituting for $\rho = a^2 r^\alpha$ in Eq. 4.13 (or equivalently $\ln\rho = 2\ln a + \alpha \ln \rho$ in Eq. 4.14), it becomes

$$\frac{kT_0}{\mu_0 m_H} \alpha = -4\pi G a^2 r^{\alpha+2}, \tag{4.15}$$

which requires $\alpha = -2$. The constant a^2 is consequently determined by the same Eq. 4.15 as

$$a^2 = \frac{kT_0}{2\pi G \mu_0 m_H}. \tag{4.16}$$

The obtained solution, $\rho(r) = a^2 r^{-2}$, is called *singular* isothermal sphere, because it is not defined at centre, being $\lim_{r\to 0} \rho(r) = +\infty$. In spite of the central density singularity, the mass contained in any spherical neighbourhood of the centre of radius ϵ is finite

$$M(\epsilon) = 4\pi a^2 \int_0^\epsilon r^{-2} r^2 \,\mathrm{d}r = 4\pi a^2 \epsilon, \quad \forall \epsilon > 0, \tag{4.17}$$

so that the real physical problem with this solution is not in the origin, but rather on the large scale, because if the material sphere has no finite boundary its mass would diverge linearly with r.

The solution for an isothermal sphere of finite central density ($\rho(0) = \rho_0 > 0$) can, as for any other polytropic sphere of arbitrary value of the exponent γ in the EOS, be obtained numerically, according to the following procedure.

[1] If $\alpha > 0$,

$$\frac{\mathrm{d}p}{\mathrm{d}r} = \frac{kT_0}{\mu_0 m_H} a^2 \alpha \rho^{\alpha-1} > 0,$$

contrary to the necessary condition for equilibrium, $\mathrm{d}p/\mathrm{d}r < 0$.

Assuming $p = c_\gamma \rho^\gamma$ with c_γ a positive constant, the equilibrium equation 4.10 becomes

$$c_\gamma \gamma \rho^{\gamma-2} \frac{d\rho}{dr} = -G\frac{M(r)}{r^2}, \tag{4.18}$$

that, to allow an standard numerical treatment, should not be transformed into a 2nd order differential equation—as done to obtain Eq. 4.11 with its further development that would lead to the so-called Lane-Emden equation[2]—but rather transformed into a set of two 1st order ODEs by the introduction of the auxiliary variable $u = GM(r)$

$$\begin{cases} c_\gamma \gamma \rho^{\gamma-2} \dfrac{d\rho}{dr} = -\dfrac{u}{r^2} \\ \\ \dfrac{du}{dr} = 4\pi G \rho r^2 \end{cases} \tag{4.19}$$

whose initial conditions are $\rho(0) = \rho_0 > 0$ and $u(0) = 0$. The condition of centrally vanishing $u(r)$ is naturally implied by the finiteness of the supposedly spatially continuous density, because

$$M(0) = \lim_{r\to 0} 4\pi \int\limits_0^r \rho(t)t^2\, dt = 4\pi \lim_{r\to 0} \rho(\xi_r)\frac{r^3}{3} = \frac{4\pi}{3}\rho_0 \lim_{r\to 0} r^3 = 0, \tag{4.20}$$

where we applied the Lagrange mean value theorem to the integral, with $0 \le \xi_r \le r$, and thanks to the continuity of $\rho(r)$ ($\lim_{r\to 0_+} \rho(\xi_r) = \rho_0$).

Once that the equilibrium condition is set in the form of the 1st order ODE system 4.19, every standard method of numerical integration of 1st order ODEs and of systems of ODEs can be applied.

Finally, we remind that the case of isothermal, finite density, sphere is obtained letting $\gamma = 1$ and $c_1 = kT_0/(\mu_0 m_H)$ in the system 4.19.

4.2.2 *Gravitating Systems out of Equilibrium*

Consider a gas in a spherically symmetric configuration of radius R and whose pressure is spatially uniform ($p = const.$). Suppose also that at the centre there is a point—like object of mass $M \gg M_g$ where M_g is the total gas mass.

Under such conditions, the Euler's equation writes

[2] J.H. Lane and R. Emden in the second half of XXIX century developed the theory of polytropic spheres at equilibrium, which joins the equilibrium condition and the Poisson's equation for the gravitational potential of a polytropic gas into a single 2nd order dimensionless equation called, after them, Lane-Emden equation.

$$\frac{\partial v_r}{\partial t} + v_r \frac{\partial v_r}{\partial r} = \frac{\mathrm{d}U}{\mathrm{d}r}, \tag{4.21}$$

where, in our case, $\mathrm{d}U/\mathrm{d}r = -(GM/r^2)$ because the gas is not self-gravitating. If the flow is stationary, the first term in the equation above can be integrated for the velocity of the generic gas particle that initially was at distance r_0 from the attraction centre with initial speed v_{r0}, giving

$$v_r(r) = \pm\sqrt{v_{r0}^2 + 2\frac{GM}{r_0}\left(\frac{r_0}{r} - 1\right)}. \tag{4.22}$$

If $v_{r0} = 0$, the minus sign in the above equation is taken, because of the attractive nature of the point mass gravity.[3]

By means of Eq. 4.22, the time needed to reach the origin starting at rest from r_0 (*free-fall* time) is obtained as

$$t_{ff}(r_0) = \int_{r_0}^{0} \frac{\mathrm{d}r}{v_r} = \frac{r_0^{3/2}}{\sqrt{2GM}} \int_0^1 \frac{\mathrm{d}x}{\sqrt{1/x - 1}}, \tag{4.23}$$

having set $x = r/r_0$. The rightmost integral in Eq. 4.23, which we denote by I, upon the substitution $x = \sin^2\theta$ transforms into

$$I = 2\int_0^{\pi/2} \sin^2\theta \mathrm{d}\theta, \tag{4.24}$$

which is solved by parts, giving $I = \pi/2$, so that, eventually,

$$t_{ff}(r_0) = \frac{\sqrt{2}}{4}\pi \frac{r_0^{3/2}}{\sqrt{GM}}. \tag{4.25}$$

Note that t_{ff} is an increasing function of r_0 so to justify as definition of free-fall time of the whole gas cloud its maximum value, $t_{ff}(R)$.

In a similar way it is possible to evaluate the free-fall time in the somehow opposite case when the spherical gas cloud, supposed uniform in density ($\rho = const.$) does not carry a compact point-like mass M at its centre or, similarly, when the gas mass M_g is overabundant compared to a possible central point-like mass, $M_g \gg M$. In such case the gas is self-gravitating and undergoes to collapse due to its own gravitation.

The way to obtain the free-fall time in this case is the same as before, but with

[3] Note that v_r is diverging for $r \to 0$.

$$\frac{dU}{dr} = -\frac{GM_g}{R^3}r, \tag{4.26}$$

so to obtain ($v_{r0} = 0$, and letting again $x = r/r_0$)

$$t_{ff}(r_0) = \frac{R^{3/2}}{\sqrt{GM_g}} \int_0^1 \frac{dx}{\sqrt{1-x^2}} = \frac{R^{3/2}}{\sqrt{GM_g}} \frac{\pi}{2}. \tag{4.27}$$

The time to reach the origin in this case of uniform density cloud is independent of r_0 and is therefore taken as global free-fall time.

In the assumption that the gas mass is equal to what was assumed to be the central point-like mass, $M_g = M$, we see that the free-fall time of Eq. 4.27 is $\sqrt{2}$ longer than that of Eq. 4.25 where $r_0 = R$ is set in the latter.

4.3 Jeans's Theory of Gravitational Instability

How stars are born is, in astrophysics, a question still partially unanswered.

That stars are originated from a diffuse gaseous medium is certain and this constitutes a sort of phase transition, but the actual modes that lead from a dilute medium to a compact state are not deeply known. As every phase transition it depends on the onset of some kind of instability which, over a certain time scale, leads the system from one equilibrium to another. A local portion of a dilute gas undergoes to a collapse instability until a sort of equilibrium is reached in an almost spherical, compact, configuration. The specific mechanism which triggers the instability is not uniquely determined and is likely different from case to case.

Surely the formation of a new star would involve processes different than those characterizing the formation of a planet. Observations indeed show that stars are formed in the so-called *cores* of giant tenuous and cold nebulae mainly composed by molecular hydrogen (citedd previously as Large Molecular Clouds, LMCs), which are irregular in shape, while planets are found around stars on almost coplanar orbits, suggesting their origin from some accretion mechanism of gas left from the parent star formation and essentially confined in the star equatorial plane. The role of gravity is undoubtedly important in both such example cases, but surely in a different way.

Let us concentrate on the case of a diffuse, large and cold extension of H_2 gas, as an LMC. The estimated kinetic temperature of these clouds are around 10 K and the matter density is very low ($\sim$100 molecules per cubic centimetre). In spite of the low density, these clouds are so spatially extended that their mass overcomes 10^5 $M_\odot$ and so, being very cold, their gravitational energy is large enough to make them intrinsically unstable against gravity. For such kind of systems it is worth studying the so-called *gravitational instability* mechanism proposed by J.H. Jeans in 1902.

Let us consider the, idealized, case of a spatially infinite extension of an ideal fluid, uniform in density and at rest. The fluid is, so, characterized by $\rho = \rho_0 > 0$, $p = p_0 > 0$, $\mathbf{v} = \mathbf{v}_0 = 0$. It is straightforward verifying that this set of constant values satisfies the governing equations (continuity, Euler's equation and isoentropic equation) if body forces are absent.

Now, let us consider a perturbation of the global equilibrium state of the fluid by additive quantities $\delta\rho(\mathbf{r};t)$, $\delta p(\mathbf{r};t)$, $\delta\mathbf{v}(\mathbf{r};t)$ such that $\rho = \rho_0 + \delta\rho$, $p = p_0 + \delta p$, $\mathbf{v} = \delta\mathbf{v}$. In order to linearize the governing equations it is required that perturbations are "small", that is $|\delta\rho/\rho_0| \ll 1$ and $|\delta p/p_0| \ll 1$. Also $\delta\mathbf{v}$ must be small respect to some typical velocity characterizing the fluid (*not* respect to $\mathbf{v}_0$ which is zero). To quantify this in the given context the request is $\|\delta\mathbf{v}\| \ll c_s$, where c_s is the fluid sound speed

$$c_s = \sqrt{\left(\frac{\partial p}{\partial \rho}\right)_{ad}}, \tag{4.28}$$

which, evaluated at time zero, gives

$$c_{s0} = \sqrt{\gamma \frac{p_0}{\rho_0}}, \tag{4.29}$$

upon the adoption of the $p \propto \rho^\gamma$ EOS with constant γ.

Inserting $\rho = \rho_0 + \delta\rho$, $p = p_0 + \delta p$, $\mathbf{v} = \delta\mathbf{v}$ in the continuity and Euler's equations, in absence of body forces, and keeping only linear terms we have

$$\begin{cases} \dfrac{\partial \delta\rho}{\partial t} + \rho_0 \nabla \cdot \delta\mathbf{v} = 0 \\ \\ \dfrac{\partial \delta\mathbf{v}}{\partial t} = -\dfrac{1}{\rho_0}\nabla \delta p. \end{cases} \tag{4.30}$$

Moreover, at first order,

$$\delta p = p(\rho_0 + \delta\rho) - p(\rho_0) = \left(\frac{\mathrm{d}p}{\mathrm{d}\rho}\right)_0 \delta\rho, \tag{4.31}$$

and so $c_{s0}^2 = \delta p/\delta\rho$. Taking the partial derivative with respect to t of both sides of the first equation in Eq. 4.30 and taking the divergence of the second equation after the substitution $\delta p = c_{s0}^2 \delta\rho$ in its right hand side, and, finally, making a side by side subtraction of the obtained second equation from the first equation, we get eventually the single scalar PDE (for simplicity we indicate c_{s0} as c_s)

$$\frac{1}{c_s^2}\frac{\partial^2}{\partial t^2}\frac{\delta\rho}{\rho_0} - \nabla^2 \frac{\delta\rho}{\rho_0} = 0, \tag{4.32}$$

which is the classical linear hyperbolic equation of wave propagation in the unknown quantity $\delta\rho/\rho_0$.

A solution of Eq. 4.32 is

$$\left(\frac{\delta\rho}{\rho_0}\right)_k = A_k e^{i(\mathbf{k}\cdot\mathbf{r}\pm\omega t)}, \tag{4.33}$$

where A_k is the amplitude of the k-wave component ($|A_k| < 1$ in the linear regime), i is the imaginary unit, $\mathbf{k}$ is the wave vector, and $\omega = 2\pi\nu$ is the angular frequency, $\nu = c_s/\lambda$ being the ordinary frequency. The vector $\mathbf{k}$, whose magnitude is $k = 2\pi/\lambda$, points in the direction of phase velocity, which is usually normal to the propagating wavefront.

It is easily seen by substitution of 4.33 in Eq. 4.32 that satisfaction of Eq. 4.32 is subjected to satisfaction of the *dispersion* relation

$$\frac{\omega^2}{c_s^2} - \mathbf{k}\cdot\mathbf{k} = 0, \tag{4.34}$$

that is $\omega^2 = c_s^2 k^2$.

Due to the linearity of the wave equation, any superposition of k-wave components is still a solution. Note that in the solution 4.33 if ω is imaginary the choice of $+$ in the argument of the exponential gives a *damping* solution (it tends to zero for t going to infinite) while the choice of $-$ gives a *growing* in time solution which, in our linearization, would lose sense when $\delta\rho/\rho_0$ overcomes 1. Anyway, in the present case of absence of body forces the dispersion relation is $\omega^2 = c_s^2 k^2 > 0$, so that the angular frequency ω is real. As a consequence, all the possible wave components of the general solution are bound in time and, indeed, the general solution is a superposition of sinusoidal oscillations. This situation corresponds to actual *stability* of the fluid upon small perturbations.

The situation changes when considering a body force in Euler's equation. In this case the evolutive equation 4.32 of density perturbations becomes

$$\frac{1}{c_s^2}\frac{\partial^2}{\partial t^2}\frac{\delta\rho}{\rho_0} - \nabla^2\frac{\delta\rho}{\rho_0} + \frac{1}{c_s^2}\nabla\cdot\frac{\delta\mathbf{F}_B}{\rho_0} = 0, \tag{4.35}$$

where $\delta\mathbf{F}_B$ is a perturbative body force (per unit volume). Let us assume a linear dependence on $\delta\rho/\rho_0$ of the body force contribution in the above equation, i.e.

$$\nabla\cdot\frac{\delta\mathbf{F}_B}{\rho_0} = a\frac{\delta\rho}{\rho_0}, \tag{4.36}$$

where $a \neq 0$ is a constant.

Upon this assumption, the plane wave function of Eq. 4.33 still satisfies the wave equation provided that the new dispersion relation

$$\frac{\omega^2}{c_s^2} - \mathbf{k} \cdot \mathbf{k} - \frac{a}{c_s^2} = 0, \tag{4.37}$$

is satisfied. Clearly, if $a \geq 0$, $\omega^2 = c_s^2 k^2 + a \geq 0$ for every k. When $a < 0$, it results in ω^2 being less, equal, or greater than zero when $k^2 < -a/c_s^2$, $k^2 = -a/c_s^2$, $k^2 > -a/c_s^2$ respectively. Denoting as $k_c^2 \equiv -a/c_s^2 > 0$ the critical (separation) value, we have that ω is imaginary for $k^2 < k_c^2$ and real for $k^2 \geq k_c^2$. As a consequence, in the general solution of Eq. 4.35 all the wave components with $k < k_c$ are "unstable" in what they correspond to one exponentially diverging in time and one exponentially vanishing in time solution, in pair. On the other side, all the large wave numbers ($k \geq k_c$) correspond to limited oscillations, as it happens in the absence of body force.

Now the question is: how is a determined in the case of the hypothetical infinite, homogeneous medium initially at rest studied by Jeans?

To answer this, we have to assume that the local body force perturbation is of gravitational origin. In a uniform density, infinitely extended medium, gravity is balanced everywhere, so every point in the medium is an equilibrium point. But a local perturbation of density causes an unbalance:

intuitively, a local $\delta\rho > 0$ (overdensity) in a point makes it an "attraction" point while the opposite happens when $\delta\rho < 0$ (underdensity).

Due to that, assuming, reasonably, a spherical symmetry in a neighbourhood of the considered point, the resulting perturbing force $\delta\mathbf{F}$ is radially oriented, naturally inwards if $\delta\rho > 0$ and outward if $\delta\rho < 0$. This can be expressed by $\delta\mathbf{F} = \pm A^2 \mathbf{e}_r$, taking $-$ (attraction) in overdensity and $+$ (repulsion) in underdensity.

Consequently

$$\nabla \cdot \frac{\delta\mathbf{F}}{\rho_0} = a\frac{\delta\rho}{\rho_0} = \pm A^2 \nabla \cdot \mathbf{e}_r = \pm A^2 \frac{2}{r}, \tag{4.38}$$

the first equality above standing for the given assumption on body force divergence-density perturbation linearity (Eq.4.36). Equation 4.38 can be satisfied only if $a < 0$: in such case, indeed, a positive $\delta\rho$ gives the right "$-$" sign in the body force perturbation divergence and a negative $\delta\rho$ gives the right "$+$" sign.

As a matter of fact, not only the sign of a is determined but also its absolute value, which simply derives by the assumed validity of the Poisson's equation that links matter density to the resulting gravitational potential. As we said above, in the case of an infinite and homogeneous fluid, $\rho = \rho_0 =$ const., the gravitational force per unit mass $\mathbf{F}_0$ is zero in every point, so that if $\mathbf{F}_0$ admits a potential function U it would be $\mathbf{F}_0 = \nabla U_0 = 0$, implying $(\nabla \cdot \nabla U)_0 = (\nabla^2 U)_0 = 0$ (satisfying Laplace equation) while the Poisson's equation would require instead $(\nabla^2 U)_0 = -4\pi G\rho_0 \neq 0$. This consideration rules out the validity of Poisson's equation in the original unperturbed situation. If, anyway, we force the assumption of its validity (Jeans's *swindle*) we can benefit from linearity of Poisson's equation to obtain in the perturbed case

$$\nabla \cdot (\mathbf{F}_0 + \delta\mathbf{F}) = -4\pi G(\rho_0 + \delta\rho) \Rightarrow \nabla \cdot \delta\mathbf{F} = -4\pi G\delta\rho, \tag{4.39}$$

leading to $a = -4\pi G\rho_0$ as value of the quantity a in Eq. 4.36 that we have previously obtained as necessarily negative on a heuristic base.

Now, on the basis of the dispersion relation in Eq. 4.37, we have two exponential solutions in time provided $k^2 < k_c^2$ (and so ω imaginary), because in this case $\omega = i|\omega|$, and so the solutions 4.33 can be written as

$$\left(\frac{\delta\rho}{\rho_0}\right)_{\pm} = \begin{cases} A_k e^{i\mathbf{k}\cdot\mathbf{r}} e^{-|\omega|t}, \\ A_k e^{i\mathbf{k}\cdot\mathbf{r}} e^{|\omega|t}, \end{cases} \tag{4.40}$$

taking in the exponential of 4.33 the plus and minus sign.

Of course, also the sum $(\delta\rho/\rho_0)_+ + (\delta\rho/\rho_0)_-$ still solves the wave equation.

The condition $k^2 < k_c^2$ is equivalent to $\lambda^2 > 4\pi^2/k_c^2$, that is $\lambda^2 > \pi\gamma k T_0/(\mu_0 m_H G\rho_0)$ because $c_s^2 = \gamma p/\rho_0$. Upon the further assumption of ideal gas, $p = k\rho T/(\mu) m_H$, the instability condition translates into a critical wavelength

$$\lambda_c = \sqrt{\frac{\pi\gamma k T_0}{\mu_0 m_H G\rho_0}}, \tag{4.41}$$

which naturally corresponds to a critical mass scale

$$M_c = \frac{4\pi}{3}\rho_0\left(\frac{\lambda_c}{4}\right)^3, \tag{4.42}$$

as the mass contained in the compression (rarefaction) region and so undergoing to gravitational instability (other authors use $\lambda_c/2$ instead of $\lambda_c/4$ in Eq. 4.42). The above expression underlines the dependence of M_c on $\rho_0^{-1/2}T^{3/2}$, marking the opposite role of gravity *against* thermal pressure in the instability.

In its limitations, Jeans's theory robustly supports a scenario where large, almost uniform clouds of gas are indeed prone to gravitational instability even if their mass density is low whenever they are cold enough. The process of local collapse likely involves different zones of the extended low temperature region leading to a sort of cascade of further "fragmentations" of the dilute medium. It is worth remembering, however, that Jeans's is a linear theory and so it loses predictive validity when the amplitude of the absolute value of the density contrast $\delta\rho/\rho_0$ approaches unity from lower values. A full study of the collapse of gaseous clouds requires thus a complete hydrodynamical and energetic treatment.

To give an astrophysical example, in the disk of the Milky Way there are several Large Molecular Clouds composed mainly of molecular hydrogen (H_2) with mass of thousands solar masses over a spatial size in the range 5–200 pc with a mass density up to 1000 particles per cubic centimetre (the average density in the solar neighbourhood is about 1 particle per cubic centimetre). The typical temperature is 10 K. If we take, as example, $T = 10$ K, $\rho_0 = 10^{-22}$ g cm^{-3} and $\gamma = 7/5$, the critical mass is $M_c \approx 14.75$ M$_\odot$, that means that almost all LMCs with this temperature and density are unstable against density perturbations.

Historical Note

The developer of the theory of gravitational instability was James Hopwood Jeans (b. 1877 in Ormskirk, UK, d. 1946 in Dorkirk, UK). Initially Jeans was interested in classical studies but later, also thanks to a very good mathematics teacher, he turned his attention to mathematics. He was particularly fascinated by numbers and their theory. At age of 19 he went to the Trinity College in Cambridge (the same where Isaac Newton worked some 250 years before) with a mathematics scholarship. Among his teachers there were A.N. Whitehead and E.T. Whittaker. Jeans was awarded an Isaac Newton Studentship in astronomy and optics, then in 1901 he was elected a Fellow of Trinity. During those times he turned his attention from number theory to physics, in both its theoretical and experimental aspects.

In 1904 he published *The dynamical theory of gases*, his first important contribution to gas and fluid dynamics. That same year Jeans was appointed a Lecturer in Mathematics at Cambridge.

In 1905 he started lectures as professor of Applied Mathematics at Princeton (USA) until 1909. After writing another book, *Theoretical Mechanics*, and after developing (together with J.W.S. Rayleigh, b. 1842 at Langford Grove, UK, d. in 1919 at Witham, UK) a theory of black-body radiation, he turned his scientific interest towards cosmology and astrophysics. In these fields he gave fundamental theoretical insights, although he was wrong when he conjectured a steady state cosmology, i.e. an universe where matter is continuously created. Another failure occurred when in a controversy with the other British scientist A. Eddington (b. 1882 in Kendal, UK, d. 1944 in Cambridge, UK) he defended his own idea that the origin of the energy released by the Sun was gravitational contraction while Eddington was right in saying that the energy source is slow nuclear fusion in the inner solar region. In spite of these two failures, Jeans achieved many important and original results, the most relevant is probably the theory of gravitational instability which explains the primary reason for contraction of interstellar dilute nebulae towards the process of star formation. Another, less known and acknowledged, great result was the deduction of the set of partial differential equations describing in a compact way the motion of stars in a gravitational field using a method derived from gas dynamics. This set of equation has been called "Jeans equations" after him.

Exercise 4.1 Show that the equilibrium equation 4.10 for a monoatomic gas embedded in a medium exerting on its boundary a pressure p_{ext} admits as first integral the quantity $2E_i - p_{ext}4\pi R^3 + \Omega = 0$, where E_i is the total internal energy of the self-gravitating fluid and Ω its gravitational energy. R is the radius of the equilibrium spherical configuration, which is assumed finite, as is its mass M.

Solution Let us write Eq. 4.10 in the form

$$\mathrm{d}p = \rho \frac{\mathrm{d}U}{\mathrm{d}r} \mathrm{d}r,$$

and multiply both its sides by $4\pi r^3$ and integrate over the whole spherical configuration, to obtain

$$\int_{p(0)}^{p(R)} 4\pi r^3 \mathrm{d}p = \int_0^R \frac{\mathrm{d}U}{\mathrm{d}r} r 4\pi \rho r^2 \mathrm{d}r. \tag{4.43}$$

The lhs integral above may be integrated by parts[4] letting $u(r) = r^3$ and $\mathrm{d}v = \mathrm{d}p$

$$\int_{p(0)}^{p(R)} 4\pi r^3 \mathrm{d}p = 4\pi \left[r^3 p \Big|_0^R - 3 \int_0^R p r^2 \mathrm{d}r \right],$$

that is

$$\int_{p(0)}^{p(R)} 4\pi r^3 \mathrm{d}p = 4\pi \left[R^3 p(R) - \lim_{r \to 0} r^3 p(r) - 3 \int_0^R p r^2 \mathrm{d}r \right]. \tag{4.44}$$

The equilibrium condition requires that at the boundary $p(R) = p_{ext}$ where p_{ext} is the pressure exerted by the external medium, so that Eq. 4.44, assuming $\lim_{r \to 0} r^3 p(r) = 0$ and accounting for the EOS $p = (\gamma - 1)\rho e$, becomes

$$\int_{p(0)}^{p(R)} 4\pi r^3 \mathrm{d}p = p_{ext} 4\pi R^3 - 3(\gamma - 1) E_i, \tag{4.45}$$

where

$$E_i = \int_0^R \rho e 4\pi r^2 \mathrm{d}r = \int_M e\, \mathrm{d}m,$$

is the total internal energy.

Inserting the above expression 4.45 into Eq. 4.43 it results

[4] Integration by parts bases on $\int_a^b u \mathrm{d}v = uv]_a^b - \int_a^b v \mathrm{d}u$.

$$p_{ext}4\pi R^3 - 3(\gamma - 1)E_i = \int\limits_M \frac{\mathrm{d}U}{\mathrm{d}r} r \mathrm{d}m,$$

and the integral in the rhs is clearly the gravitational energy, Ω,[5] finally leading to

$$3(\gamma - 1)E_i - p_{ext}4\pi R^3 + \Omega = 0. \tag{4.46}$$

For an equilibrium monoatomic gas ($\gamma = 5/3$) the above relation reduces to $2E_i - p_{ext}4\pi R^3 + \Omega = 0$. If the sphere is in vacuum ($p_{ext} = 0$), or extends to infinite the relation simplifies to $2E_i + \Omega = 0$.

4.4 Further Readings

This chapter presented fundaments of potential theory and of harmonic functions theory, with application to the specific Newtonian gravitational fields. Although self-consistent and easily to understand without additional readings, the reader who wants to know more about potential theory, harmonic functions and the, related, field of partial differential equations, would find useful the references [1–5].

References

1. Gilbarg, D., Trudinger, N.S.: Elliptic Equations of Second Order. Springer, New York (1977)
2. Kellogg, O.D.: Foundations of Potential Theory. Dover, New York (1953)
3. Mc Robert, T.M.: Spherical Harmonics. Pergamon Press, Oxford (1967)
4. Mikhailov, V.P.: Partial Differential Equations. MIR Publishers, Moscow (1978)
5. Vladimirov, V.S.: Equations of Mathematical Physics. MIR Publishers, Moscow (1984)

[5] The integrand is indeed the work per unit mass done against gravitational (radial) force.

Chapter 5
Classical Potential Theory

In this chapter we look for approximate solutions of Poisson's equation for regular distributions of matter in the absence of spatial symmetry—a common scenario in fields from astrophysics to electrostatics. This is classically approached via series expansions (specifically, the multipole expansion), whose coefficients can be determined by appropriate use of harmonic analysis. Ultimately, this chapter contains the fundamentals of what is generally referred to as "classical potential theory".

5.1 Fundamental Solution of Laplace Equation

We saw in Chap. 1 that in $\mathbb{R}^n$

$$\begin{cases} \nabla \cdot \nabla \dfrac{1}{r^{n-2}} = 0, \text{ if } n = 1 \text{or } n > 2, \\ \\ \nabla \cdot \nabla \ln \dfrac{1}{r} = 0, \quad \text{if } n = 2, \end{cases} \tag{5.1}$$

with $r \equiv \sqrt{\sum_{i=1}^{n} x_i^2} > 0$.

Let us consider the generic point $\mathbf{r}' \in \mathbb{R}^n$ as "pole" and define $\tilde{\mathbf{r}} = \mathbf{r} - \mathbf{r}'$, so that $\tilde{r} \equiv \|\tilde{\mathbf{r}}\|$. Given a function $f(\tilde{r})$, letting $\widetilde{\nabla}$ be the formal nabla operator acting on the coordinates of $\tilde{\mathbf{r}}$, and recalling that $\widetilde{\nabla} \cdot \mathbf{e}_{\tilde{r}} = \dfrac{n-1}{\tilde{r}}$, it results

$$\begin{aligned} \widetilde{\nabla}^2 f(\tilde{r}) &= \widetilde{\nabla} \cdot \widetilde{\nabla} f(\tilde{r}) = \widetilde{\nabla} \cdot f'(\tilde{r})\mathbf{e}_{\tilde{r}} = f'(\tilde{r})\widetilde{\nabla} \cdot \mathbf{e}_{\tilde{r}} + \mathbf{e}_{\tilde{r}} \cdot \widetilde{\nabla} f'(\tilde{r}) = \\ &= \frac{n-1}{\tilde{r}} f'(\tilde{r}) + \mathbf{e}_{\tilde{r}} \cdot f''(\tilde{r})\mathbf{e}_{\tilde{r}} = \frac{n-1}{\tilde{r}} f'(\tilde{r}) + f''(\tilde{r}), \end{aligned} \tag{5.2}$$

R. A. Capuzzo Dolcetta, *Classical Newtonian Gravity and Potential Theory*,
UNITEXT for Physics, https://doi.org/10.1007/978-3-032-13036-5_5

where the prime stands for derivative with respect to $\tilde{r}$.

Consequently, the Laplace partial differential equation $\widetilde{\nabla}^2 f(\tilde{r}) = 0$ reduces to an ordinary differential equation

$$\frac{n-1}{\tilde{r}} f'(\tilde{r}) + f''(\tilde{r}) = 0, \tag{5.3}$$

which, in the case $n = 3$, writes also as

$$\frac{1}{\tilde{r}^2} \frac{\mathrm{d}}{\mathrm{d}\tilde{r}} \left[\tilde{r}^2 f'(\tilde{r}) \right] = 0. \tag{5.4}$$

The differential equation 5.3 is solved by letting $g(\tilde{r}) = f'(\tilde{r})$, so that the 2nd order equation transforms into the 1st order system

$$\begin{cases} f'(\tilde{r}) = g(\tilde{r}) \\ \dfrac{n-1}{\tilde{r}} g(\tilde{r}) + g'(\tilde{r}) = 0, \end{cases} \tag{5.5}$$

which has ($\forall n$) the trivial solution $f(\tilde{r}) = const.$ and the solution

$$g(\tilde{r}) = c_0 \tilde{r}^{1-n}, \tag{5.6}$$

so that $f(\tilde{r})$ comes out by a quadrature of $g(\tilde{r})$, to get

$$\begin{cases} f(\tilde{r}) = c_0 \dfrac{\tilde{r}^{2-n}}{2-n} + c_1, \text{ if } n = 1 \text{ or } n > 2, \\ f(\tilde{r}) = c_0 \ln \tilde{r} + c_1, \quad \text{if } n = 2. \end{cases} \tag{5.7}$$

Letting, in Eq. 5.7,

$$\begin{cases} c_0 = -G, c_1 = 0, \text{for } n = 1 \text{ or } n > 2, \\ c_0 = G, c_1 = 0, \text{for } n = 2, \end{cases} \tag{5.8}$$

one obtains the so-called *fundamental solution* of the Laplace equation (or *fundamental harmonic*[1]):

[1] In $\mathbb{R}$ every linear function $f(x) = ax + b$ is harmonic.

$$\begin{cases} \Gamma_n(\|\mathbf{r}-\mathbf{r}'\|) = -G\dfrac{1}{(2-n)\|\mathbf{r}-\mathbf{r}'\|^{n-2}}, & \text{for } n=1 \text{ or } n>2, \\ \Gamma_2(\|\mathbf{r}-\mathbf{r}'\|) = G\ln\|\mathbf{r}-\mathbf{r}'\|, & \text{for } n=2, \end{cases} \tag{5.9}$$

hence the definition of volume potential as a *convolution* product

$$U_C(\mathbf{r}) = \underbrace{\int\cdots\int_C}_{n} \rho(\mathbf{r}')\Gamma_n(\|\mathbf{r}-\mathbf{r}'\|)\,\mathrm{d}^n\mathbf{r}' \equiv \rho * \Gamma_n, \tag{5.10}$$

where C is an nD matter domain, and the operator "$*$" represents the convolution product.

In the case $n = 3$ we obtain the already known (see Sect. 2.2) expression for the potential

$$U_C(\mathbf{r}) = G\iiint_C \frac{\rho(\mathbf{r}')}{\|\mathbf{r}-\mathbf{r}'\|}\,\mathrm{d}^3\mathbf{r}', \tag{5.11}$$

with $\rho(\mathbf{r}')$ being a bounded and integrable function in the domain C.

If $n = 2$ the convolution gives

$$U_C(\mathbf{r}) = G\iint_C \rho(\mathbf{r}')\ln\|\mathbf{r}-\mathbf{r}'\|\,\mathrm{d}^2\mathbf{r}', \tag{5.12}$$

and so

$$\nabla\cdot\nabla U_C = -2\pi G\rho, \tag{5.13}$$

which is the Poisson's equation in 2D.

In the general case of n dimensions it is

$$\nabla\cdot\nabla U_C = -n\omega_n G\rho, \tag{5.14}$$

where $n\omega_n = \sigma_n$ represents the area of the nD unit ball whose volume measures ω_n.

5.2 Harmonic Functions

Let S represent the ball of unitary radius. On its surface, consider the spherical triangle with curved sides abc with angles at corners $\widehat{A}, \widehat{B}, \widehat{C}$ (opposed, respectively, to the curved sides a, b, c, obtained as intersection of the spherical surface with three planes through the sphere centre (Fig. 5.1)).

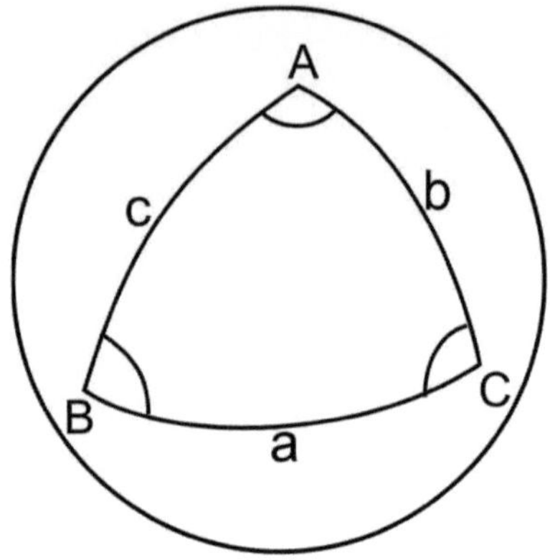

Fig. 5.1 The spherical triangle ABC. Source https://commons.wikimedia.org/wiki/File:Spherical_trigonometry_legendre.svg (author: PeterMercator)

For this kind of triangle the following theorems of spherical trigonometry hold:

Theorem 5.1 (Sine Theorem)

$$\frac{\sin \widehat{A}}{\sin a} = \frac{\sin \widehat{B}}{\sin b} = \frac{\sin \widehat{C}}{\sin c}. \tag{5.15}$$

Theorem 5.2 (Cosine Theorem)

$$\begin{cases} \cos a = \cos b \cos c + \sin b \sin c \cos \widehat{A}, \\ \cos b = \cos a \cos c + \sin a \sin c \cos \widehat{B}, \\ \cos c = \cos a \cos b + \sin a \sin b \cos \widehat{C}, \end{cases} \tag{5.16}$$

$$\begin{cases} \cos \widehat{A} = -\cos \widehat{B} \cos \widehat{C} + \sin \widehat{B} \sin \widehat{C} \cos a, \\ \cos \widehat{B} = -\cos \widehat{A} \cos \widehat{C} + \sin \widehat{A} \sin \widehat{C} \cos b, \\ \cos \widehat{C} = -\cos \widehat{A} \cos \widehat{B} + \sin \widehat{A} \sin \widehat{B} \cos c. \end{cases} \tag{5.17}$$

Note that Theorems 5.1 and 5.2 are the equivalent, in spherical trigonometry, of the sine theorem and of the cosine (Carnot's) theorem in the plane,[2] which give, respectively

$$\frac{\sin \widehat{A}}{a} = \frac{\sin \widehat{B}}{b} = \frac{\sin \widehat{C}}{c}, \tag{5.18}$$

and

$$\begin{cases} a^2 = b^2 + c^2 - 2bc \cos \widehat{A}, \\ b^2 = a^2 + c^2 - 2ac \cos \widehat{B}, \\ c^2 = a^2 + b^2 - 2ab \cos \widehat{C}. \end{cases} \tag{5.19}$$

[2] Carnot's theorem is the generalization of Pythagorean theorem.

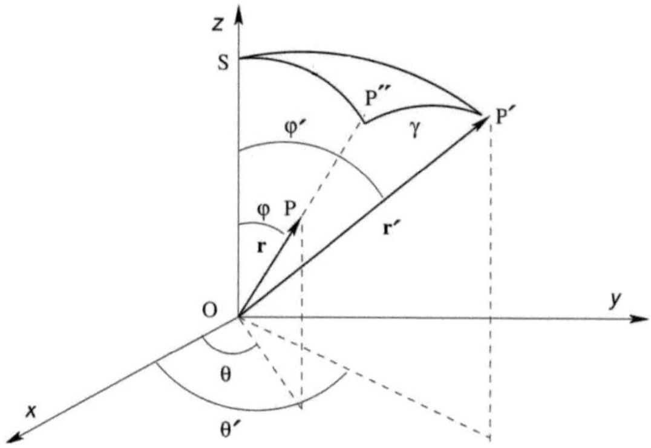

Fig. 5.2 The spherical triangle $P'P''S$

Now, let $\mathbf{r}$ and $\mathbf{r}'$ be the position vectors of two arbitrary points P and P' of unit mass. Setting the gravitational constant $G = 1$, the Newtonian potential in P as due to the presence of P' (and vice versa) is given by

$$\Gamma_3(\|\mathbf{r}-\mathbf{r}'\|) = \frac{1}{\|\mathbf{r}-\mathbf{r}'\|}, \qquad \text{with} \quad \|\mathbf{r}-\mathbf{r}'\| = \sqrt{r^2 + r'^2 - 2rr'\cos\gamma}, \tag{5.20}$$

after defining the angle $\gamma = \widehat{\mathbf{rr}'}$.

so to give

$$\frac{1}{\|\mathbf{r}-\mathbf{r}'\|} = \frac{1}{\sqrt{r^2 + r'^2 - 2rr'\cos\gamma}} = \frac{1}{r'}\frac{1}{\sqrt{\left(\frac{r}{r'}\right)^2 + 1 - 2\frac{r}{r'}\cos\gamma}}. \tag{5.21}$$

Resorting to spherical polar coordinates such that $\mathbf{r} = (r, \theta, \varphi)$ and $\mathbf{r}' = (r', \theta', \varphi')$, and applying the cosine Theorem 5.2 to the $P'P''S$ spherical triangle drawn on the spherical surface of radius r' (see Fig. 5.2), it follows that

$$\cos\widehat{P'P''} = \cos\widehat{\mathbf{rr}'} \equiv \cos\gamma = \cos\varphi\cos\varphi' + \sin\varphi\sin\varphi'\cos(\theta - \theta'). \tag{5.22}$$

Inserting the above expression for $\cos\gamma$ into Eq. 5.21 provides for $1/\|\mathbf{r}-\mathbf{r}'\|$ a complicated function of $r, r', \theta, \theta', \varphi, \varphi'$. Our aim is to check what kind of approximation we can introduce to make the expression of $1/\|\mathbf{r}-\mathbf{r}'\|$ easier to deal with.

Letting in Eq. 5.21 $\mu \equiv \dfrac{r}{r'}$ and $u \equiv \cos\gamma$, it can be written as

$$\frac{1}{\|\mathbf{r}-\mathbf{r}'\|} = \frac{1}{r'}\frac{1}{\sqrt{\mu^2 + 1 - 2\mu u}}. \tag{5.23}$$

Our purpose is now that of obtaining a proper series expansion in μ of the above function.

Note that the function $(1+x)^n$, where n is a generic real number, admits a development in series of x around $x = 0$, such that

(*i*) it converges for $|x| < 1$, if $n \le -1$;
(*ii*) it converges for $-1 < x \le 1$, if $-1 < n < 0$;
(*iii*) it converges for $|x| \le 1$, if $n > 0$ and $n \notin \mathbb{Z}^+$
(*iv*) it reduces to the usual Newton binomial expansion, if $n \in \mathbb{Z}^+$, and is valid for every x.

In general, the formal expression is (note the alternate change of sign)

$$(1 \pm x)^n = 1 \pm nx + \frac{n(n-1)}{2}x^2 \pm \cdots + (\pm 1)^k \frac{n(n-1)\cdots(n-k+1)}{k!}x^k + \\ + \cdots\cdots \equiv \sum_{k=0}^{\infty} \left(\!\!\binom{n}{k}\!\!\right) (\pm 1)^k x^k, \tag{5.24}$$

where in the above sum the sign + is chosen if k is even and $-$ if k is odd, and where we have introduced a double round parenthesis symbol as a generalization of the usual binomial coefficient (this latter is defined when both n and k are non negative integers, while here n is not necessarily a positive integer). Actually, for $n \in \mathbb{Z}^+$, we have, by our definition

$$\left(\!\!\binom{n}{k}\!\!\right) \equiv \frac{n(n-1)\cdots(n-k+1)}{k!} = \frac{n!}{k!(n-k)!} \equiv \binom{n}{k}. \tag{5.25}$$

Going back to Eq. 5.23, the function

$$\frac{1}{\sqrt{\mu^2 + 1 - 2\mu u}} = \left[1 + (\mu^2 - 2u\mu)\right]^{-1/2}, \tag{5.26}$$

is of the type $(1+x)^n$, once assumed $x \equiv \mu(\mu - 2u)$ and $n = -1/2$.

Consequently,

$$\left(\!\!\binom{n}{k}\!\!\right) = \left(\!\!\binom{-\frac{1}{2}}{k}\!\!\right) = \frac{\overbrace{-\frac{1}{2}\left(-\frac{3}{2}\right)\left(-\frac{5}{2}\right)\cdots\left(\frac{1}{2}-k\right)}^{k\ factors}}{k(k-1)\cdots\cdots\cdots 2\cdot 1} = \\ = \frac{-\frac{1}{2}\left(-\frac{3}{2}\right)\left(-\frac{5}{2}\right)\cdots\left(-\left[\frac{2k-1}{2}\right]\right)}{k(k-1)\cdots\cdots\cdots 2\cdot 1} =$$

$$= (-1)^k \frac{1}{2^k} \frac{1 \cdot 3 \cdot 5 \cdots (2k-1)}{k(k-1) \cdot \cdots \cdot 2 \cdot 1} = \tag{5.27}$$

$$= (-1)^k \frac{1 \cdot 3 \cdot 5 \cdots (2k-1)}{2k \cdot 2(k-1) \cdots 4 \cdot 2} =$$

$$= (-1)^k \frac{\prod_{l=1}^{k} (2l-1)}{\prod_{l=1}^{k} 2l},$$

where the final ratio is that between the product of the first k odd integers to that of the first k even integers.

As consequence of what above, we have

$$\frac{1}{\sqrt{\mu^2 + 1 - 2\mu u}} = \sum_{k=0}^{\infty} \left(\binom{-\frac{1}{2}}{k} \right) \mu^k (\mu - 2u)^k =$$
$$= \sum_{k=0}^{\infty} (-1)^k \frac{\prod_{l=1}^{k} (2l-1)}{\prod_{l=1}^{k} 2l} \mu^k (\mu - 2u)^k. \tag{5.28}$$

The above series is *not* a simple power series but, rather, a polynomial series in μ, convergent for $0 \le \mu \le u + \sqrt{u^2+1}$ that, together with the allowed variation of u, says that the series above converges pointwise for $\mu \le 1$, $\forall u \in]-1, 1[$. Moreover, uniform convergence stands $\forall [-a, a] \in]-1, 1[$.

The series 5.28 can be also written in the form

$$\frac{1}{\sqrt{\mu^2 + 1 - 2\mu u}} = \sum_{k=0}^{\infty} \left(\binom{-\frac{1}{2}}{k} \right) \mu^k \sum_{l=0}^{k} \binom{k}{l} \mu^l (-1)^{k-l} (2u)^{k-l} =$$
$$= \sum_{k=0}^{\infty} \left(\binom{-\frac{1}{2}}{k} \right) \sum_{l=0}^{k} \mu^{k+l} V_{l,k}, \tag{5.29}$$

where we set

$$V_{l,k} = (-1)^{k-l} \binom{k}{l} (2u)^{k-l}. \tag{5.30}$$

The coefficient of the generic μ^m is given by the sum of the coefficients of all the terms μ^{k+l}, with k and l variables under the constraint $k + l = m$ with $l \le k$

Table 5.1 Values of k and l allowed upon the conditions $k + l = m$ and $l \leq k$

	k	l
	m	0
	$m-1$	1
	$m-2$	2
	$\vdots$	$\vdots$
if m is even	$m/2$	$m/2$
if m is odd	$[m/2]+1$	$[m/2]$

according to Table 5.1. Accordingly, the coefficient of μ^m is given by

$$\left(\!\binom{-\frac{1}{2}}{m}\!\right) V_{0,m} + \left(\!\binom{-\frac{1}{2}}{m-1}\!\right) V_{1,m-1} + \left(\!\binom{-\frac{1}{2}}{m-2}\!\right) V_{2,m-2} + \cdots + \left(\!\binom{-\frac{1}{2}}{K}\!\right) V_{L,K}, \tag{5.31}$$

where in the last term $L = K = m/2$ for m even, while $L = [m/2]$, $K = [m/2]+1$ for m odd.[3]

The expression above comprises $[m/2] + 1$ alternating terms and it is actually a polynomial of degree m in the variable u, that we call $P_m(u)$, whose extended expression comes from the below development[4]

$$\frac{1}{\sqrt{\mu^2 + 1 - 2\mu u}} = \sum_{k=0}^{\infty} \left(\!\binom{-\frac{1}{2}}{k}\!\right) \sum_{l=0}^{k} \mu^{k+l} V_{l,k} =$$

$$= \sum_{m=0}^{\infty} \left[\sum_{\substack{l \leqslant k \\ l+k=m}} \left(\!\binom{-\frac{1}{2}}{k}\!\right) \underbrace{(-1)^{k-l} \binom{k}{l} (2u)^{k-l}}_{V_{l,k}} \right] \mu^m =$$

$$= \sum_{m=0}^{\infty} \left[\sum_{\substack{l \leqslant k \\ l+k=m}} \left(\!\binom{-\frac{1}{2}}{m-l}\!\right) (-1)^{m-2l} \binom{m-l}{l} 2^{m-2l} u^{m-2l} \right] \mu^m =$$

$$= \sum_{m=0}^{\infty} \left[\sum_{l=0}^{[\frac{m}{2}]} (-1)^{m-l} \underbrace{\frac{1 \cdot 3 \cdots 2(m-l)-1)}{2 \cdot 4 \cdots 2(m-l)}}_{2^{m-l}(m-l)!} (-1)^{m-2l} \frac{(m-l)!}{l!(m-2l)!} 2^{m-2l} u^{m-2l} \right] \mu^m =$$

[3] $[\frac{m}{2}]$ is the integer part of $m/2$.

[4] Note that $(-1)^{2m-3l} = (-1)^l$.

$$= \sum_{m=0}^{\infty} \left[\sum_{l=0}^{[\frac{m}{2}]} \frac{1 \cdot 3 \cdots (2(m-l)-1)}{\cancel{2^{m-l}(m-l)!}} (-1)^l \cancel{2^{m-l}(m-l)!} \frac{1}{2^l} \frac{1}{l!(m-2l)!} u^{m-2l} \right] \mu^m =$$

$$= \sum_{m=0}^{\infty} \left[\sum_{l=0}^{[\frac{m}{2}]} \frac{1 \cdot 3 \cdots (2(m-l)-1)}{2^l \, l! \, (m-2l)!} (-1)^l \, u^{m-2l} \right] \mu^m =$$

$$= \sum_{m=0}^{\infty} P_m(u) \mu^m, \tag{5.32}$$

where we set

$$P_m(u) = \sum_{l=0}^{[\frac{m}{2}]} \frac{1 \cdot 3 \cdots (2(m-l)-1)}{2^l \, l! \, (m-2l)!} (-1)^l \, u^{m-2l}. \tag{5.33}$$

A more compact form of the expression above is

$$P_m(u) = \sum_{l=0}^{[\frac{m}{2}]} (-1)^l \frac{\prod_{k=1}^{m-l} (2k-1)}{2^l \, l! \, (m-2l)!} u^{m-2l}. \tag{5.34}$$

Another, alternative, expression for $P_m(u)$ is

$$P_m(u) = \sum_{l=0}^{[\frac{m}{2}]} \frac{(2(m-l))!}{2^m l!(m-l)!(m-2l)!} (-1)^l \, u^{m-2l}. \tag{5.35}$$

$P_m(u)$ is called *Legendre polynomial* of degree m, and it has a relevant importance in mathematical physics. Due to that

$$\frac{1}{\sqrt{\mu^2 + 1 - 2\mu u}} = \sum_{m=0}^{\infty} P_m(u) \mu^m, \tag{5.36}$$

the function $(\mu^2 + 1 - 2\mu u)^{-1/2}$ is called "generating" function of the Legendre polynomials.

Note 15
Note that, because

$$\begin{cases} m - 2l & \text{is even when } m \text{ is even}, \\ m - 2l & \text{is odd when } m \text{ is odd}, \end{cases}$$

(continued)

Note 15 (continued)
and because $(-u)^{m-2l} = (-1)^{m-2l} u^{m-2l}$, it results

$$\boxed{P_m(-u) = (-1)^m P_m(u)} \tag{5.37}$$

The terms in the Legendre polynomials are of alternating sign, and the highest power of u has always a positive sign. $P_m(u)$ contains only even powers of u, if m is even, and only odd powers of u, if m is odd. This resumes as

$$\boxed{\begin{array}{ll} P_{2m}(u) = P_{2m}(-u), & \text{so } P_{2m}(u) \text{ is an even function} \\ P_{2m+1}(u) = -P_{2m+1}(u), & \text{so } P_{2m+1}(u) \text{ is an odd function} \end{array}} \tag{5.38}$$

Finally, the series expansion of the fundamental harmonic

$$\frac{1}{\|\mathbf{r}' - \mathbf{r}\|} = \frac{1}{r'} \frac{1}{\sqrt{\mu^2 + 1 - 2\mu u}} = \frac{1}{r'} \sum_{m=0}^{\infty} P_m(u)\mu^m = \sum_{m=0}^{\infty} P_m(u) \frac{r^m}{r'^{\,m+1}} \tag{5.39}$$

is such that the following theorem holds:

Theorem 5.3 *The series*

$$\frac{1}{\|\mathbf{r} - \mathbf{r}'\|} = \frac{1}{r'} \sum_{m=0}^{\infty} P_m(u)\mu^m, \tag{5.40}$$

converges for

$$\mu = \frac{r}{r'} < 1, \qquad u = \cos\gamma \in [-1, 1] \Longleftrightarrow \gamma \in [0, \pi].$$

Proof Using the known Euler's formula

$$u = \cos\gamma = \frac{\mathrm{e}^{i\gamma} + \mathrm{e}^{-i\gamma}}{2},$$

it results

$$\frac{1}{\sqrt{\mu^2 + 1 - 2\mu u}} = \frac{1}{\sqrt{\left(1 - \mu \mathrm{e}^{i\gamma}\right)\left(1 - \mu \mathrm{e}^{-i\gamma}\right)}} =$$

$$= \left(1 - \mu e^{i\gamma}\right)^{-\frac{1}{2}} \left(1 - \mu e^{-i\gamma}\right)^{-\frac{1}{2}} = \tag{5.41}$$
$$= \sum_{k=0}^{\infty} \alpha_k e^{ki\gamma} \mu^k \sum_{k=0}^{\infty} \alpha_k e^{-ki\gamma} \mu^k,$$

where we expanded in series of $x = \mu e^{i\gamma}$ and $y = \mu e^{-i\gamma}$ the two expressions $(1 - x)^{-\frac{1}{2}}$ and $(1 - y)^{-\frac{1}{2}}$. The two series are converging $\forall \gamma \in \mathbb{R}$ if $|\mu| < 1$, so that the original series with coefficients $P_m(u)$ converges for $\mu < 1$.

The expression 5.40 allows writing $P_m(u)$ as a trigonometric polynomial:

$$\begin{aligned} P_m(u) &= 2\alpha_0\alpha_m \cos(m\gamma) + 2\alpha_1\alpha_{m-1} \cos[(m-2)\gamma]+ \\ &\quad + 2\alpha_2\alpha_{m-2} \cos[(m-4)\gamma] + \cdots\cdots . \end{aligned} \tag{5.42}$$

Actually, $P_m(u)$ is obtained by the sum of trigonometric terms in (5.41) with indices that give m as sum. For example, the second addend above comes this way

$$\begin{aligned} 2\alpha_1\alpha_{m-1} \cos[(m-2)\gamma] &= \alpha_1\alpha_{m-1} e^{i\gamma} e^{-i(m-1)\gamma} + \alpha_{m-1}\alpha_1 e^{i(m-1)\gamma} e^{-i\gamma} = \\ &= \alpha_1\alpha_{m-1} \left(e^{i\gamma(2-m)} + e^{-i\gamma(2-m)}\right) = \\ &= 2\alpha_1\alpha_{m-1} \cos[(2-m)\gamma], \end{aligned} \tag{5.43}$$

and, thus, for the $(k + 1)$th term we have

$$2\alpha_k\alpha_{m-k} \cos[(2k - m)\gamma]. \tag{5.44}$$

Finally,

$$P_m(u) = \sum_{k=0}^{[\frac{m}{2}]} 2\alpha_k\alpha_{m-k} \cos[(2k - m)\gamma], \tag{5.45}$$

and, so,

$$P_m(1) = \sum_{k=0}^{[\frac{m}{2}]} 2\alpha_k\alpha_{m-k}. \tag{5.46}$$

Moreover, $\alpha_k > 0 \; \forall k$, because

$$(1 - x)^n = \sum_{k=0}^{\infty} (-1)^k \left(\binom{-\frac{1}{2}}{k}\right) x^k = \sum_{k=0}^{\infty} \underbrace{(-1)^{2k} \frac{1 \cdot 3 \cdot 5 \cdots (2k-1)}{2 \cdot 4 \cdot 6 \cdots 2k}}_{\alpha_k > 0} x^k. \tag{5.47}$$

Being all coefficients positive, $P_m(u)$ takes its maximum value in $u = \cos\gamma = 1$; actually

$$P_m(u) = \sum_{k=0}^{[\frac{m}{2}]} 2\alpha_k\alpha_{m-k}\cos[(2k-m)\gamma] \leq \sum_{k=0}^{[\frac{m}{2}]} 2\alpha_k\alpha_{m-k} = P_m(1). \tag{5.48}$$

Finally, let us note that, if m is even, in the expression of $P_m(u)$ the constant term $2\alpha_{\frac{m}{2}}^2$ appears, otherwise there is the term $2\alpha_{\frac{m-1}{2}}\alpha_{\frac{m+1}{2}}\cos\gamma$.

5.3 Legendre Polynomials

Recalling the expressions 5.34 and 5.35 for the Legendre polynomials, it is possible to compute their coefficients.

Actually, Legendre polynomials with $m = 0, \cdots, 5$ are explicitly given by

$$\begin{aligned}
P_0(u) &= 1,\\
P_1(u) &= u,\\
P_2(u) &= \frac{1}{2}(3u^2-1),\\
P_3(u) &= \frac{1\cdot3\cdot5}{3!}u^3 - \frac{1\cdot3}{2}u = \frac{1}{2}(5u^3-3u),\\
P_4(u) &= \frac{1}{8}(35u^4-30u^2+3),\\
P_5(u) &= \frac{1}{8}(63u^5-70u^3+15u).
\end{aligned} \tag{5.49}$$

The zeros of such polynomials are

$$\begin{aligned}
P_1(u) = 0 &\Longleftrightarrow u_1 = 0,\\
P_2(u) = 0 &\Longleftrightarrow u_{1,2} = \pm\sqrt{\tfrac{1}{3}},\\
P_3(u) = 0 &\Longleftrightarrow u_1 = 0,\ u_{2,3} = \pm\sqrt{\tfrac{3}{5}},\\
P_4(u) = 0 &\Longleftrightarrow u_{1,2} = \pm\sqrt{\tfrac{15+2\sqrt{30}}{35}},\ u_{3,4} = \pm\sqrt{\tfrac{15-2\sqrt{30}}{35}},\\
P_5(u) = 0 &\Longleftrightarrow u_1 = 0,\ u_{2,3} = \pm\sqrt{\tfrac{35+2\sqrt{70}}{63}},\ u_{4,5} = \pm\sqrt{\tfrac{35-2\sqrt{70}}{63}}.
\end{aligned}$$

Plots of Legendre polynomials up to the 5th order are shown in Fig 5.3.

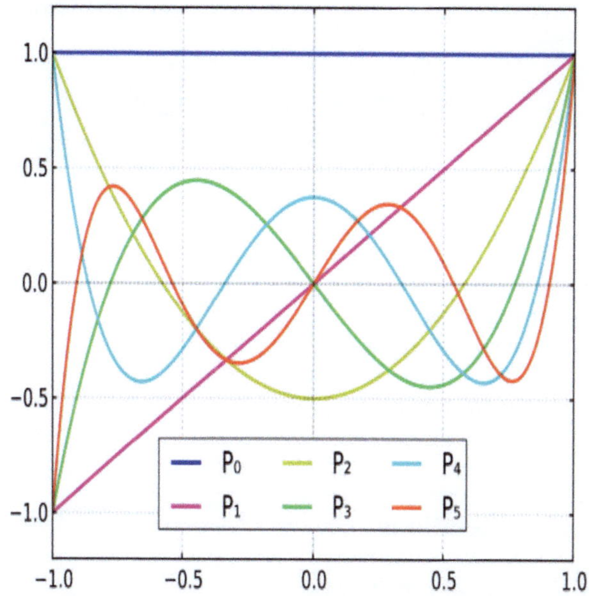

Fig. 5.3 Legendre polynomials $P_m(u)$ for $0 \leq m \leq 5$. Source https://commons.wikimedia.org/w/index.php?curid=32145761 (author: Geek3)

5.3.1 Some Properties of Legendre Polynomials

The Legendre polynomials of odd index vanish in $u = 0$ because they contain the linear term u as lowest power of u. On the contrary, $P_m(u)$ with even index m do not vanish because they contain a nonzero constant term. To summarize

$$\begin{aligned} P_{2k}(0) &\neq 0, \\ P_{2k+1}(0) &= 0, \end{aligned} \tag{5.50}$$

for every non-negative integer k.

Additionally, note that

$$\begin{aligned} P_m(1) &= 1, \\ P_m(-1) &= (-1)^m, \end{aligned} \tag{5.51}$$

for every non-negative integer m, as we now easily show.

Actually, letting $u = 1$, it results

$$\frac{1}{\sqrt{\mu^2 + 1 - 2\mu}} = \sum_{m=0}^{\infty} P_m(1)\mu^m. \tag{5.52}$$

For $|\mu| < 1$, another series expansion is possible (geometric series in μ)

$$\frac{1}{\sqrt{\mu^2+1-2\mu}} = \frac{1}{1-\mu} = \sum_{m=0}^{\infty} u^m, \tag{5.53}$$

so that, by comparison of 5.52 and 5.53 it results

$$P_m(1) = 1, \tag{5.54}$$

for every non-negative integer m.

Additionally, letting $u = 1$ in Eq. 5.37 we get

$$P_m(-1) = (-1)^m P_m(1) = (-1)^m. \tag{5.55}$$

5.4 Recurrence Relations

5.4.1 First Recurrence Relation

Because the series $\sum_{m=0}^{\infty} P_m(u)\mu^m$ is uniformly convergent for $|\mu| \le 1$ over every compact interval of u contained in $]-1, 1[$, it is term-by-term differentiable with respect to μ and the obtained series converges to the derivative (with respect to μ) of $\dfrac{1}{\sqrt{\mu^2+1-2\mu u}}$.

From this, it results that

$$\begin{aligned}\frac{\mathrm{d}}{\mathrm{d}\mu}\frac{1}{\sqrt{\mu^2+1-2\mu u}} &= \frac{u-\mu}{\left[\mu^2+1-2\mu u\right]^{3/2}} = \sum_{m=0}^{\infty} P_m(u)\frac{\mathrm{d}\mu^m}{\mathrm{d}\mu} = \\ &= \sum_{m=0}^{\infty} P_m(u) m \mu^{m-1},\end{aligned} \tag{5.56}$$

which, after some manipulation, leads to

$$\frac{u-\mu}{\sqrt{\mu^2+1-2\mu u}}\,\frac{1}{\mu^2+1-2\mu u} = \sum_{m=0}^{\infty} P_m(u) m \mu^{m-1}, \tag{5.57}$$

and to

$$(u-\mu)\sum_{m=0}^{\infty} P_m(u)\mu^m = (\mu^2+1-2u\mu)\sum_{m=0}^{\infty} P_m(u) m \mu^{m-1}, \tag{5.58}$$

and, carrying the coefficients into the summation symbol

$$\sum_{m=0}^{\infty} P_m(u)\left(u\mu^m - \mu^{m+1}\right) = \sum_{m=0}^{\infty} P_m(u)m\left(\mu^{m+1} + \mu^{m-1} - 2u\mu^m\right). \tag{5.59}$$

Applying to this latter expression the polynomial identity principle (coefficients of same power of μ must be equal on both sides) we finally have this 1st *recurrence* relation

$$\boxed{(m+1)P_{m+1}(u) - (2m+1)uP_m(u) + mP_{m-1}(u) = 0, \quad \forall m \geq 1.} \tag{5.60}$$

5.4.2 Second Recurrence Relation

By differentiation with respect to u of the generating function we have

$$\frac{\mathrm{d}}{\mathrm{d}u}\frac{1}{\sqrt{\mu^2+1-2\mu u}} = \frac{\mu}{\left(\mu^2+1-2\mu u\right)^{3/2}}, \tag{5.61}$$

The function in the rhs of the above equation admits a series expansion, that we formally write as

$$\frac{\mu}{\left(\mu^2+1-2\mu u\right)^{3/2}} = \sum_{m=0}^{\infty} \widetilde{P}_m(u)\mu^m. \tag{5.62}$$

We now verify that

$$\widetilde{P}_m(u) = \frac{\mathrm{d}}{\mathrm{d}u}P_m(u). \tag{5.63}$$

For this purpose, we integrate with respect to u from 0 to generic u both sides of 5.62, obtaining

$$\begin{aligned} \int_0^u \frac{\mu}{\left(\mu^2+1-2\mu u\right)^{3/2}}\,\mathrm{d}u &= \frac{1}{\sqrt{\mu^2+1-2\mu u}} - \frac{1}{\sqrt{\mu^2+1}} = \\ = \sum_{m=0}^{\infty} P_m(u)\mu^m - \sum_{m=0}^{\infty} P_m(0)\mu^m &= \sum_{m=0}^{\infty} \mu^m \int_0^u \widetilde{P}_m(u)\,\mathrm{d}u. \end{aligned} \tag{5.64}$$

The above relation implies

$$\int_0^u \widetilde{P}_m(u)\,\mathrm{d}u = P_m(u) - P_m(0), \tag{5.65}$$

and so

$$\widetilde{P}_m(u) = \frac{\mathrm{d}}{\mathrm{d}u} P_m(u) \equiv P'_m(u), \tag{5.66}$$

which finally yields to

$$\frac{\mathrm{d}}{\mathrm{d}u} \frac{1}{\sqrt{\mu^2 + 1 - 2\mu u}} = \frac{\mu}{(\mu^2 + 1 - 2\mu u)^{3/2}} = \sum_{m=0}^{\infty} P'_m(u)\mu^m. \tag{5.67}$$

A multiplication of both the 2nd and 3rd sides of relation 5.67 by $u - \mu$ gives the following sequence of equivalent equalities

$$\begin{aligned}
\mu \frac{(u-\mu)}{\left(\mu^2 + 1 - 2u\mu\right)^{\frac{3}{2}}} &= (u-\mu) \sum_{m=0}^{\infty} \mu^m P'_m(u), \\
\mu \frac{\mathrm{d}}{\mathrm{d}\mu} \frac{1}{\sqrt{\mu^2 + 1 - 2u\mu}} &= (u-\mu) \sum_{m=0}^{\infty} \mu^m P'_m(u), \\
\mu \frac{\mathrm{d}}{\mathrm{d}\mu} \sum_{m=0}^{\infty} \mu^m P_m(u) &= (u-\mu) \sum_{m=0}^{\infty} \mu^m P'_m(u), \\
\mu \sum_{m=0}^{\infty} m\mu^{m-1} P_m(u) &= \sum_{m=0}^{\infty} P'_m(u)(u\mu^m - \mu^{m+1}),
\end{aligned} \tag{5.68}$$

hence, the second recurrence relation:

$$\boxed{P_m(u)m = u\, P'_m(u) - P'_{m-1}(u), \quad \forall m \geq 1.} \tag{5.69}$$

5.5 The Legendre Differential Equation

The first and second recurrence relations, Eqs. 5.60 and 5.69, lead to the differential equation satisfied by the generic Legendre polynomial $P_m(u)$.

Differentiation with respect to u of Eq. 5.60 gives

$$(m+1)P'_{m+1}(u) - (2m+1)P_m(u) - (2m+1)uP'_m(u) + mP'_{m-1}(u) = 0, \tag{5.70}$$

which, upon elimination of $P'_{m-1}(u)$ via 5.69, becomes

$$(m+1)P'_{m+1}(u) - (2m+1)P_m(u) - (2m+1)uP'_m(u)$$

$$+m\left[uP'_m(u) - mP_m(u)\right] = 0. \tag{5.71}$$

If we write the above expression reducing by 1 the indices ($m+1 \to m$) we get

$$\begin{aligned} mP'_m(u) - [2(m-1)+1]P_{m-1}(u) - [2(m-1)+1]uP'_{m-1}(u) + \\ +(m-1)[uP'_{m-1}(u) - (m-1)P_{m-1}(u)] = 0. \end{aligned} \tag{5.72}$$

By collecting coefficients, one finds

$$\begin{aligned} mP'_m(u) + uP'_{m-1}(u)\overbrace{[m-1-2(m-1)-1]}^{-m} + \\ - P_{m-1}(u)\overbrace{[2(m-1)+1+(m-1)^2]}^{m^2} = 0, \end{aligned} \tag{5.73}$$

which reduces to

$$P'_m(u) - uP'_{m-1}(u) - mP_{m-1}(u) = 0, \tag{5.74}$$

which, upon eliminating $P'_{m-1}(u)$ by means of (5.69), becomes

$$(1-u^2)P'_m(u) + muP_m(u) - mP_{m-1}(u) = 0, \tag{5.75}$$

that, upon differentiation with respect to u, leads to

$$\frac{\mathrm{d}}{\mathrm{d}u}[(1-u^2)P'_m(u)] + mP_m(u) + muP'_m(u) - mP'_{m-1}(u) = 0, \tag{5.76}$$

which, after a further elimination of $P'_{m-1}(u)$ with (5.69), leads to see that $P_m(u)$ satisfies the 2nd order Legendre differential equation

$$\boxed{\frac{\mathrm{d}}{\mathrm{d}u}[(1-u^2)P'_m(u)] + m(m+1)P_m(u) = 0.} \tag{5.77}$$

A third recurrence relation can be obtained by relating $P'_m(u)$, $P'_{m-2}(u)$, and $P_{m-1}(u)$ by means of Eqs. 5.69 and 5.74 (the reader is suggested to verify it, as exercise)

$$\boxed{P'_m(u) - P'_{m-2}(u) - (2m-1)P_{m-1}(u) = 0, \quad \forall m \geq 2}. \tag{5.78}$$

5.6 Orthogonality of Legendre Polynomials $P_m(u)$

Let $P_m(u)$ and $P_n(u)$ be two Legendre polynomials of arbitrary degrees m and n.

Theorem 5.4 *The polynomials $P_m(u)$ and $P_n(u)$ are orthogonal in the interval $[-1, 1]$. More specifically, we have*

$$(P_m(u), P_n(u)) \equiv \int_{-1}^{1} P_m(u)\, P_n(u)\, \mathrm{d}u = \frac{2}{2m+1}\delta_{mn}, \tag{5.79}$$

where

$$\delta_{mn} = \begin{cases} 0, & m \neq n \\ 1, & m=n \end{cases} \tag{5.80}$$

is the usual Kronecker symbol.

Proof Let us consider first $m \neq n$. Polynomials $P_m(u)$ and $P_n(u)$ satisfy the Legendre equation (5.77), i.e.

$$\frac{\mathrm{d}}{\mathrm{d}u}[(1-u^2)P'_m(u)] + m(m+1)P_m(u) = 0,$$
$$\frac{\mathrm{d}}{\mathrm{d}u}[(1-u^2)P'_n(u)] + n(n+1)P_n(u) = 0,$$

so that by multiplication of both sides of the first of the above equation by $P_n(u)$ and of the second by $P_m(u)$, respectively, followed by a subtraction of the second equation from the first, we get

$$\begin{aligned} &\frac{\mathrm{d}}{\mathrm{d}u}\left\{(1-u^2)\left[P'_m(u)P_n(u) - P'_n(u)P_m(u)\right]\right\} = \\ &= [n(n+1) - m(m+1)]\, P_m(u)P_n(u). \end{aligned} \tag{5.81}$$

Note that the left hand side of the equation above is obtained by simple operations, as follows

$$\begin{aligned} &P_n(u)\frac{\mathrm{d}}{\mathrm{d}u}[(1-u^2)P'_m(u)] - P_m(u)\frac{\mathrm{d}}{\mathrm{d}u}[(1-u^2)P'_n(u)] = \\ &= \frac{\mathrm{d}}{\mathrm{d}u}[(1-u^2)P'_m(u)P_n(u)] - \frac{\mathrm{d}}{\mathrm{d}u}[(1-u^2)P'_n(u)P_m(u)] = \\ &= \frac{\mathrm{d}}{\mathrm{d}u}\left\{(1-u^2)\left[P'_m P_n(u) - P'_n P_m(u)\right]\right\}. \end{aligned}$$

Integrating with respect to u between -1 and 1 both sides of Eq. 5.81 it results

$$\Big[(1-u^2)(P_m' P_n - P_n' P_m)\Big]_{-1}^{1} = [m(m+1) - n(n+1)] \int_{-1}^{1} P_m(u) P_n(u)\, \mathrm{d}u,$$

where the lhs is, clearly, zero, because $P_m(u), P_n(u) \in \mathcal{C}^{\infty}([-1,1])$. Then, it follows that

$$\int_{-1}^{1} P_m(u) P_n(u)\, \mathrm{d}u = 0,$$

as we wanted to show.

Let us now treat the case $m = n$. Consider the generating function of $P_m(u)$ and square both sides of its series expansion, to obtain

$$\frac{1}{\mu^2 + 1 - 2\mu u} = \left(\sum_{m=0}^{\infty} P_m(u)\mu^m\right)^2 = \sum_{m=0}^{\infty} P_m(u)\mu^m \sum_{n=0}^{\infty} P_n(u)\mu^n =$$
$$= \sum_{m=0}^{\infty}\sum_{n=0}^{\infty} P_m(u)P_n(u)\mu^m\mu^n,$$

which, upon integration over u in $[-1, 1]$, yields

$$\int_{-1}^{1} \frac{1}{\mu^2 + 1 - 2\mu u}\, \mathrm{d}u = \sum_{m=0}^{\infty}\sum_{n=0}^{\infty} \left(\int_{-1}^{1} P_m(u)P_n(u)\, \mathrm{d}u\right) \mu^{m+n}.$$

For $n \neq m$, all the integrals at rhs are zeros, as we saw above, the only surviving integral is for $n = m$, so that

$$-\frac{1}{2\mu} \ln \frac{(1-\mu)^2}{(1+\mu)^2} = \sum_{m=0}^{\infty} \mu^{2m} \int_{-1}^{1} P_m^2(u)\, \mathrm{d}u.$$

In the above equation, we have

$$-\frac{1}{2\mu} \ln \frac{(1-\mu)^2}{(1+\mu)^2} = \frac{1}{\mu} \ln \frac{1+\mu}{1-\mu} = \frac{1}{\mu} 2 \sum_{m=0}^{\infty} \frac{\mu^{2m+1}}{2m+1} = 2 \sum_{m=0}^{\infty} \frac{\mu^{2m}}{2m+1}$$

having expanded in series of μ the second expression from left. Finally, we have

$$2\sum_{m=0}^{\infty}\frac{\mu^{2m}}{2m+1}=\sum_{m=0}^{\infty}\mu^{2m}\int_{-1}^{1}P_m^2(u)\,\mathrm{d}u$$

that, for the identity by series principle, implies

$$\int_{-1}^{1}P_m^2(u)\,\mathrm{d}u=\frac{2}{2m+1}.$$

Finally, for every (m,n) positive integer pair, it is

$$(P_m(u),\,P_n(u))=\frac{2}{2m+1}\,\delta_{mn}, \tag{5.82}$$

as we wanted to demonstrate.

From the above theorem it results the following

Lemma 5.5 *The set of polynomials* $\overline{P}_m(u)=\left\{\sqrt{\frac{2m+1}{2}}\,P_m(u)\right\}$ *is orthonormal in the interval* $[-1,1]$. *Moreover, because the set of* $P_m(u)$ *polynomials is dense in the Hilbert space* $L^2([-1,1])$, *the set* $\overline{P}_m(u)$ *represents an orthonormal base of the Hilbert space* $L^2([-1,1])$.

5.7 Expansion of a Function in Series of Legendre Polynomials

Consider a function $f(u)$ defined over the real interval $[-1,1]$. It can be expanded in series of Legendre polynomials adopting the orthonormal function base $\{\overline{P}_m(u)\}=\{\sqrt{(2m+1)/2}\,P_m(u)\}$, if $f(u)$ satisfies the following *Dirichlet conditions*:

(*1*) $f(u)\in\mathcal{C}^0([-1,1])$ except for a finite number of discontinuity points;
(*2*) $f(u)$ has a finite number of minima and maxima in the $[-1,1]$ interval.

Upon the above constraints, as further sufficient (but not necessary) condition, the series

$$\sum_{m=0}^{\infty}(f,\overline{P}_m)\overline{P}_m(u), \tag{5.83}$$

converges in every $u \in]-1, 1[$ to:

(*i*) $f(u)$, if f is continuous in u;

(*ii*) to $\dfrac{f(u+0)+f(u-0)}{2}$, if u is one of the points in the finite set of discontinuity points for $f(u)$.[5]

The series is written as

$$f(u) = \sum_{m=0}^{\infty} c_m \overline{P}_m(u), \tag{5.84}$$

with

$$c_m = \left(f, \overline{P}_m(u)\right) = \int_{-1}^{1} f(u)\,\overline{P}_m(u)\,\mathrm{d}u. \tag{5.85}$$

Indeed, if the series converges to $f(u)$ and it is integrable term by term, it follows

$$\left(f, \overline{P}_n\right) = \int_{-1}^{1} f(u)\,\overline{P}_n(u)\,\mathrm{d}u = \sum_{m=0}^{\infty} c_m \int_{-1}^{1} \overline{P}_m(u)\overline{P}_n(u)\,\mathrm{d}u = c_n, \tag{5.86}$$

due to $\overline{P}_m(u)$ orthonormality.

Exercise 5.1 Expand the *Heaviside* step function, defined in the interval $[-1, 1]$ as

$$f(u) = \begin{cases} 0, & -1 \le u < 0, \\ 1, & 0 \le u \le 1, \end{cases}$$

in series of Legendre polynomials (Fig. 5.4).

Solution The expansion is possible because $f(u)$ is continuous in the $[-1, 1]$ interval, excluding the origin, so that (for every $m \ge 1$)

$$c_m = \int_{-1}^{1} f(u)\,\overline{P}_m(u)\,\mathrm{d}u = \int_{0}^{1} \overline{P}_m(u)\,\mathrm{d}u =$$

$$= \sqrt{\frac{2m+1}{2}}\,\frac{1}{2m+1} \int_{0}^{1} \left[P'_{m+1}(u) - P'_{m-1}(u)\right]\,\mathrm{d}u =$$

[5] Here we define $f(u \pm c) \equiv \lim_{c \to 0^+} f(u \pm c)$.

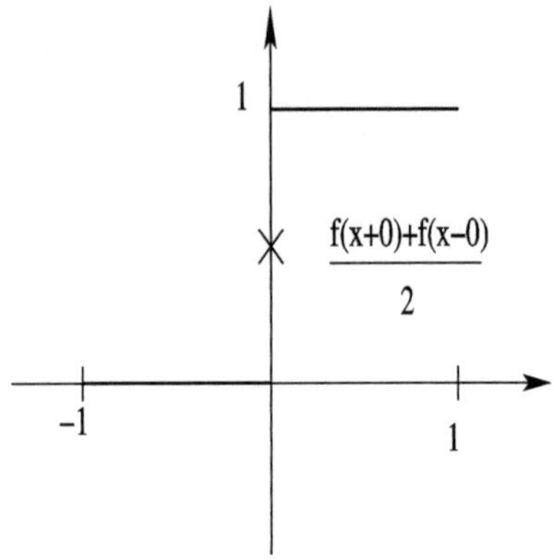

Fig. 5.4 Sketch of the Heaviside function restricted to the $[-1, 1]$ interval of x

$$
\begin{aligned}
&= \sqrt{\frac{2m+1}{2}}\frac{1}{2m+1}\left[P_{m+1}(1) - P_{m+1}(0) - P_{m-1}(1) + P_{m-1}(0)\right] = \\
&= \sqrt{\frac{2m+1}{2}}\frac{1}{2m+1}\left[-P_{m+1}(0) + P_{m-1}(0)\right] = \\
&= \sqrt{\frac{2m+1}{2}}\frac{1}{2m+1}\left[P_{m-1}(0) - P_{m+1}(0)\right],
\end{aligned}
$$

where we used the relation 5.78 and that $P_k(1) = 1, \forall k$. We have, finally

$$
c_m = \begin{cases}
\sqrt{2}/2, & \text{if } m = 0 \\
0, & \text{if } m \text{ is even} \\
\sqrt{\frac{2m+1}{2}}\frac{1}{2m+1}\left[(-1)^{\frac{m-1}{2}}\frac{1\cdot 3\cdots(m-2)}{2^{\frac{m-1}{2}}\left(\frac{m-1}{2}\right)!} - (-1)^{\frac{m+1}{2}}\frac{1\cdot 3\cdots m}{2^{\frac{m+1}{2}}\left(\frac{m+1}{2}\right)!}\right], & \text{if } m \text{ is odd}
\end{cases}
$$

5.8 Rodrigues Formula

Proposition 5.1 *The generic Legendre polynomial, $P_m(u)$, can be written as*

$$
P_m(u) = \frac{1}{2^m\, m!}\frac{\mathrm{d}^m}{\mathrm{d}u^m}\left(u^2 - 1\right)^m, \tag{5.87}
$$

expression that constitutes the Rodrigues formula.

Proof Recalling Eqs. 5.33 and 5.34 we have

$$P_m(u) = \sum_{l=0}^{[\frac{m}{2}]}(-1)^l \frac{\prod_{k=1}^{m-l}(2k-1)}{2^l\, l!\,(m-2l)!}\, u^{m-2l} =$$

$$= \frac{1\cdot 3\cdots(2m-1)}{m!}\, u^m - \frac{1\cdot 3\cdots(2m-3)}{2(m-2)!}\, u^{m-2} + \frac{1\cdot 3\cdots(2m-5)}{2^2\, 2\,(m-4)!}\, u^{m-4} - \cdots$$

$$= \frac{1\cdot 3\cdots(2m-1)}{m!}\Big[u^m - \frac{m!}{2(m-2)!(2m-1)}\, u^{m-2} +$$

$$+ \frac{m!}{2^2\, 2(m-4)!(2m-1)(2m-3)}\, u^{m-4} - \cdots\Big] =$$

$$= \frac{1}{m!}\prod_{k=1}^{m}(2k-1)\left[u^m - \frac{m(m-1)}{2(2m-1)}\, u^{m-2} + \frac{m(m-1)(m-2)(m-3)}{2^2\, 2(2m-1)(2m-3)}\, u^{m-4} - \cdots\right].$$

By m consecutive integrations over the limits 0 and u of $P_m(u)$ and of the finite product here above, we obtain

$$\int P_m(u)\, \mathrm{d}^m u \equiv \underbrace{\int \mathrm{d}u \int \mathrm{d}u \cdots \int \mathrm{d}u\, P_m(u)}_{m\ \ times} =$$

$$= \frac{1}{m!}\prod_{k=1}^{m}(2k-1)\Big[\frac{u^{2m}}{(m+1)(m+2)\cdots 2m} +$$

$$- \frac{m(m-1)}{2(2m-1)}\, \frac{u^{2m-2}}{(m-2+1)(m-2+2)\cdots(2m-2)} +$$

$$+ \frac{m(m-1)(m-2)(m-3)}{2^2 2(2m-1)(2m-3)}\, \frac{u^{2m-4}}{(m-3)(m-2)\cdots(2m-4)} - \cdots\Big] =$$

$$= \frac{1}{(2m)!}\prod_{k=1}^{m}(2k-1)\Big[u^{2m} - \frac{(2m)!}{2(2m-1)}\, \frac{u^{2m-2}}{(2m-2)!} +$$

$$+ \frac{(2m)!}{2^2 2(2m-1)(2m-3)}\, \frac{u^{2m-4}}{(2m-4)!} - \cdots\Big] =$$

$$= \frac{1}{(2m)!}\prod_{k=1}^{m}(2k-1)\left[u^{2m} - m u^{2m-2} + \frac{m(m-1)}{2} u^{2m-4} - \cdots\right] =$$

$$= \frac{1}{(2m)!}\prod_{k=1}^{m}(2k-1)\sum_{l=0}^{m}(-1)^l\binom{m}{l}(u^2)^{m-l} =$$

$$= \frac{1}{(2m)!}\prod_{k=1}^{m}(2k-1)\left(u^2-1\right)^m = \frac{1}{2^m m!}\left(u^2-1\right)^m .$$

Now, by m differentiations of the first left side and of the last right side of the above series of relations, we eventually get the Rodrigues formula

$$P_m(u) = \frac{1}{2^m m!}\frac{\mathrm{d}^m}{\mathrm{d}u^m}\left(u^2-1\right)^m . \tag{5.88}$$

Exercise 5.2 Show that the function

$$y_m(u) = k\,\frac{\mathrm{d}^m}{\mathrm{d}u^m}(u^2-1)^m,$$

satisfies the Legendre equation 5.77, and that $k = \dfrac{1}{2^m\, m!}$ if $y_m(1) = 1$.

Solution Let $x(u) = k(u^2-1)^m$; it follows that $\dfrac{\mathrm{d}x}{\mathrm{d}u} = 2kmu(u^2-1)^{m-1}$, and, multiplying by (u^2-1) both sides, it results

$$(u^2-1)\frac{\mathrm{d}x}{\mathrm{d}u} = 2kmu(u^2-1)^m = 2mux.$$

Making $m+1$ differentiations of the lhs and rhs

$$\frac{\mathrm{d}^{m+1}}{\mathrm{d}u^{m+1}}\left[(u^2-1)\frac{\mathrm{d}x}{\mathrm{d}u}\right] = 2m\frac{\mathrm{d}^{m+1}}{\mathrm{d}u^{m+1}}(ux).$$

Applying to the above expression the relation

$$\frac{\mathrm{d}^{m+1}}{\mathrm{d}u^{m+1}}(f\cdot g) = \sum_{k=0}^{m+1}\binom{m+1}{k} f^{(k)}\, g^{(m+1-k)},$$

which is valid for every $m+1$ times differentiable functions $f(u)$ and $g(u)$, we have

$$\frac{\mathrm{d}^{m+1}}{\mathrm{d}u^{m+1}}\left[(u^2-1)\frac{\mathrm{d}x}{\mathrm{d}u}\right] = (u^2-1)\frac{\mathrm{d}^{m+2}x}{\mathrm{d}u^{m+2}} + (m+1)\,2u\frac{\mathrm{d}^{m+1}x}{\mathrm{d}u^{m+1}} + \frac{m(m+1)}{2}\,2\,\frac{\mathrm{d}^m x}{\mathrm{d}u^m} =$$

$$= 2mu\frac{\mathrm{d}^{m+1}x}{\mathrm{d}u^{m+1}} + 2m(m+1)\frac{\mathrm{d}^m x}{\mathrm{d}u^m},$$

which reduces to

$$(u^2-1)\frac{\mathrm{d}^{m+2}x}{\mathrm{d}u^{m+2}}+2u\frac{\mathrm{d}^{m+1}x}{\mathrm{d}u^{m+1}}-m(m+1)\frac{\mathrm{d}^m x}{\mathrm{d}u^m}=0.$$

Letting

$$y_m=\frac{\mathrm{d}^m x}{\mathrm{d}u^m}$$

the above equation assumes the form

$$(u^2-1)\frac{\mathrm{d}^2 y_m}{\mathrm{d}u^2}+2u\frac{\mathrm{d}y_m}{\mathrm{d}u}-m(m+1)y_m=0,$$

and, collecting terms, finally

$$\frac{\mathrm{d}}{\mathrm{d}u}\left[(u^2-1)\frac{\mathrm{d}y_m}{\mathrm{d}u}\right]-m(m+1)y_m=0,$$

as we wanted to prove.

The proof that $k=\dfrac{1}{2^m\,m!}$ if $y_m(1)=1$ is left to the reader.

Exercise 5.3 Using the Rodrigues formula, show that $P_m(u)$ has m real and distinct zeroes in the $(-1,1)$ interval. Show also that, by application of Rolle's theorem, it follows that

(*a*) the zeroes of $P_{2m}(u)$ are symmetric with respect to the origin;
(*b*) the zeroes of $P_{2m+1}(u)$ are symmetric with respect to the origin and $P_{2m+1}(0)=0$.

Solution Left to the reader.

5.9 Harmonic and Homogeneous Polynomials

Let us remind the series expansion of the inverse of the Euclidean distance between two points $\mathbf{r}\equiv(x,y,z)$ and $\mathbf{r}'\equiv(x',y',z')$

$$\frac{1}{\|\mathbf{r}-\mathbf{r}'\|}=\frac{1}{r'}\sum_{m=0}^{\infty}P_m(u)\mu^m=\sum_{m=0}^{\infty}P_m(u)\mu^m\frac{r^m}{r'^{\,m+1}}\equiv\sum_{m=0}^{\infty}H_m(x,y,z), \tag{5.89}$$

where $H_m(x,y,z)$ is a homogeneous polynomial of degree m in x,y,z, when taking x',y',z' as parameters. Indeed, recalling that

$$\mu\equiv\frac{r}{r'},\qquad u\equiv\cos\gamma=\frac{\mathbf{r}\cdot\mathbf{r}'}{rr'}=\frac{xx'+yy'+zz'}{rr'}, \tag{5.90}$$

where

$$r = \sqrt{x^2 + y^2 + z^2}, \quad r' = \sqrt{x'^2 + y'^2 + z'^2}, \tag{5.91}$$

it results that the function

$$H_m = P_m(u)\frac{r^m}{r'^{m+1}}, \tag{5.92}$$

depends upon r through $u^m r^m$, $u^{m-2}r^m$, $u^{m-4}r^m$, etc., that is through r^0, r^2, r^4, etc., i.e. only through powers of r with exponents $\leq m$. Consequently, $H_m(x, y, z)$ is a homogeneous polynomial of mth degree[6] being, in its turn, u a homogeneous function of degree 0 in x, y, z. Finally,

$$H_m(x, y, z) = \sum_{\substack{(i,j,k) \\ i+j+k=m}} a_{ijk}\, x^i y^j z^k, \tag{5.93}$$

that implies that the quantity $\dfrac{1}{\|\mathbf{r} - \mathbf{r}'\|}$ admits a power series expansion[7]

$$\frac{1}{\|\mathbf{r} - \mathbf{r}'\|} \equiv \sum_{m=0}^{\infty} H_m(x, y, z) = \sum_{m=0}^{\infty} \sum_{\substack{(i,j,k) \\ i+j+k=m}} a_{ijk}\, x^i y^j z^k. \tag{5.94}$$

Because the series 5.93 converges to $\dfrac{1}{\|\mathbf{r} - \mathbf{r}'\|}$ in a neighbourhood of the origin, the function $\dfrac{1}{\|\mathbf{r} - \mathbf{r}'\|}$ is *analytic*[8] in every point $\mathbf{r} \neq \mathbf{r}'$.

In conclusion, the function $1/\|\mathbf{r} - \mathbf{r}'\|$ is a harmonic function which has an expansion in terms of homogeneous polynomials, equivalent to an expansion in power series convergent in a neighbourhood of the origin. This means that the series 5.93 is term by term differentiable, and thus

$$\nabla^2 \frac{1}{\|\mathbf{r} - \mathbf{r}'\|} = 0 = \sum_{m=0}^{\infty} \nabla^2 H_m(x, y, z). \tag{5.95}$$

A power series cannot converge to zero in a domain containing the origin if its coefficients are not all zero, that is

---

[6] All the addends are of the type $\dfrac{(xx' + yy' + zz')^{m-2k}}{r'^{\,2m-2k+1}} r^{2k}$.

[7] The expansion is of the type $\sum a_{ijk}(x - x_0)^i (y - y_0)^j (z - z_0)^k$ with $x_0 = y_0 = z_0 = 0$.

[8] In what it has a convergent power series expansion.

Table 5.2 Harmonic polynomials of the first 3 orders

m	$H_m(x, y, z)$
0	$1/r'$
1	$(xx' + yy' + zz')/r'^3$
2	$\frac{1}{2r'^5}\left[x'^2(2x^2 - y^2 - z^2) + y'^2(2y^2 - z^2 - x^2) + z'^2(2z^2 - x^2 - y^2) + \right.$ $\left. +6y'z'yz + 6x'y'xy + 6x'z'xz\right]$

Table 5.3 All the possible spherical harmonics (s.h.) of first 3 orders

m	Solid spherical harmonics	Number of s. h.
0	1	1
1	x, y, z	3
2	$2x^2 - y^2 - z^2, 2y^2 - z^2 - x^2, 2z^2 - x^2 - y^2, yz, xy, xz$	6

$$\nabla^2 H_m(x, y, z) = 0, \qquad \forall m. \tag{5.96}$$

The above conditions imply that *all* the H_m polynomials are harmonic: they are called "solid spherical harmonics".

Table 5.2 gives the expressions of the $H_m(x, y, z)$ for $m = 0, 1, 2$.

The spherical harmonics H_m depend upon x', y', z' as parameters, and the H_m are harmonics for any value of parameters, so that every single addend in the polynomial is also a spherical harmonic. This means that the solid spherical harmonics in the case $0 \le m \le 2$ are those of Table 5.3.

Note that the 2nd order ($m = 2$) harmonics in Table 5.3 are not all independent; actually, any of the first three of them can be written as linear combination of the two remaining. For instance, the first one can be written as follows

$$2x^2 - y^2 - z^2 = a(2y^2 - z^2 - x^2) + b(2z^2 - x^2 - y^2), \qquad \text{with } a = b = -1\,, \tag{5.97}$$

reducing to 5 the number of linearly independent spherical harmonics of 2nd order. This, together with the fact that the number of independent spherical harmonics of order $m = 0$ and $m = 1$ is, respectively, 1 and 3, suggests that, in general, the number of independent spherical harmonics of arbitrary order m would be $2m + 1$. This conjecture results to be true, as we now check. First, we note that the generic homogeneous polynomial of degree m can be written as

$$F_m(x, y, z) = \sum_{ijk} a_{ijk} x^i y^j z^k\,, \qquad \text{where} \begin{cases} i + j + k = m, \\ i, j, k = 0, 1, \ldots, m, \end{cases} \tag{5.98}$$

where a_{ijk} are $\frac{1}{2}(m+1)(m+2)$ constants, as we now show. Actually, the constraint $i+j+k = m$ implies that—choosing arbitrarily j and k among their $m+1$ possible

Table 5.4 For every value of j the table reports in the other columns the allowed values of k, the corresponding number of choices and the constrained values of i

j	k	Number of choices	$i = m - j - k$
0	$0, 1, \ldots, m$	$m + 1$	$m, m - 1, \ldots, 0$
1	$0, 1, \ldots, m - 1$	m	$m - 1, m - 2, \ldots, 0$
2	$0, 1, \ldots, m - 2$	$m - 1$	$m - 2, m - 3, \ldots, 0$
$\vdots$	$\vdots$	$\vdots$	$\vdots$
$m - 1$	0,1	2	1,0
m	0	1	0

values $(k+j \le m)$—i, which is always given by $i = m - j - k$, assumes the values reported in the following Table 5.4

From this table it is clear that the total number of choices is the sum of the values in the 3rd column, $1 + 2 + \cdots + (m - 1) + m + (m + 1)$, that is the arithmetic progression

$$a + (a + d) + (a + 2d) + \cdots + (a + md), \tag{5.99}$$

whose sum is

$$\sum_{k=0}^{m}(a + kd) = (m + 1)a + \frac{1}{2}(m + 1)md. \tag{5.100}$$

In our case $a = d = 1$ so to give

$$\sum_{k=0}^{m}(1 + k) = (m + 1) + \frac{1}{2}(m + 1)m = \frac{1}{2}(m + 1)(2 + m). \tag{5.101}$$

When the polynomial F_m other than homogeneous is also harmonic, it should be such that

$$\nabla^2 F_m = 0 = \sum_{ijk} a_{ijk}\left[i(i-1)x^{i-2}y^j z^k + j(j-1)x^i y^{j-2}z^k + k(k-1)x^i y^j z^{k-2}\right]. \tag{5.102}$$

$\nabla^2 F_m$ is, in its turn, a homogeneous polynomial of degree $m - 2$. It is so possible writing

$$\nabla^2 F_m = \sum_{ijk} A_{ijk}x^i y^j z^k = 0\,, \qquad i + j + k = m - 2, \tag{5.103}$$

where the coefficients A_{ijk} are linear combinations of the a_{ijk} and are such that all the $m(m-1)/2$ coefficients A_{ijk} are zeros. This implies the existence of $m(m-1)/2$

linear relations among the $(m+1)(m+2)/2$ coefficients a_{ik}. As a conclusion, a homogeneous and harmonic polynomial of degree m has $(m+1)(m+2)/2-(m-1)m/2 = 2m+1$ arbitrary coefficients. This statement corresponds to thesis of the following

Theorem 5.6 *The number of linearly independent homogeneous and harmonic polynomials of degree m is* $2m+1$.

Obviously, any of the independent homogeneous and harmonic polynomial can be obtained by making one of the $2m+1$ possible choices for the coefficients. The most general homogeneous and harmonic polynomial would be given by the expression[9]

$$F_m(x, y, z) = \sum_{k=1}^{2m+1} \alpha_k F_m^k(x, y, z), \tag{5.104}$$

where the F_m^k functions are the $2m+1$ independent homogeneous and harmonic polynomials.

5.10 Surface Spherical Harmonics

When considering spherical polar coordinates (r, θ, φ), as defined in Sect. 1.5.1, the generic order m spherical harmonic $F_m(x, y, z)$ is transformed into a function of spherical polar coordinates as follows

$$\begin{aligned} F_m(x, y, z) &= F_m(r\sin\varphi\cos\theta, r\sin\varphi\sin\theta, r\cos\varphi) = \\ &= \sum_{(ijk)=0}^{m} a_{ijk} r^{i+j+k} (\sin\varphi\cos\theta)^i (\sin\varphi\sin\theta)^j (\cos\varphi)^k = \\ &= r^m \sum_{(ijk)=0}^{m} a_{ijk} (\sin\varphi\cos\theta)^i (\sin\varphi\sin\theta)^j (\cos\varphi)^k = \\ &= r^m F_m(\sin\varphi\cos\theta, \sin\varphi\sin\theta, \cos\varphi) \equiv r^m S_m(\theta, \varphi), \end{aligned} \tag{5.105}$$

where $S_m(\theta, \varphi) \equiv F_m(\sin\varphi\cos\theta, \sin\varphi\sin\theta, \cos\varphi)$ is a homogeneous and harmonic polynomial in $(\sin\varphi\cos\theta, \sin\varphi\sin\theta, \cos\varphi)$, called *surface spherical harmonic* of order m, defined on the unitary radius sphere. As the solid spherical harmonics, also the linearly independent surface spherical harmonics of order m are in number of $2m+1$.

[9] Note that here k is a superscript index and not an exponent.

In spherical polars, the Laplacian $\nabla^2 F_m$ takes the form

$$\nabla^2 F_m = \frac{1}{r^2}\frac{\partial}{\partial r}\left(r^2\frac{\partial F_m}{\partial r}\right) + \frac{1}{r^2 \sin^2\varphi}\frac{\partial^2 F_m}{\partial \theta^2} + \frac{1}{r^2 \sin\varphi}\frac{\partial}{\partial \varphi}\left(\sin\varphi\frac{\partial F_m}{\partial \varphi}\right) =$$
$$= \frac{1}{r^2}\frac{\partial}{\partial r}\left(r^2 m r^{m-1} S_m\right) + \frac{r^m}{r^2 \sin^2\varphi}\frac{\partial^2 S_m}{\partial \theta^2} + \frac{r^m}{r^2 \sin\varphi}\frac{\partial}{\partial \varphi}\left(\sin\varphi\frac{\partial S_m}{\partial \varphi}\right) =$$
$$= m(m+1)S_m r^{m-2} + \frac{r^{m-2}}{\sin^2\varphi}\frac{\partial^2 S_m}{\partial \theta^2} + \frac{r^{m-2}}{\sin\varphi}\frac{\partial}{\partial \varphi}\left(\sin\varphi\frac{\partial S_m}{\partial \varphi}\right). \tag{5.106}$$

The above expression, valid for every $r > 0$, implies the following equation to be satisfied by the corresponding surface harmonic $S_m(\theta, \varphi)$

$$(m+1)m S_m(\theta, \varphi) + \frac{1}{\sin^2\varphi}\frac{\partial^2 S_m}{\partial \theta^2} + \frac{1}{\sin\varphi}\frac{\partial}{\partial \varphi}\left(\sin\varphi\frac{\partial S_m}{\partial \varphi}\right) = 0. \tag{5.107}$$

Letting $\cos\varphi = t$, the equation above can be written as

$$(m+1)m S_m(\theta, \varphi) + \frac{1}{1-t^2}\frac{\partial^2 S_m}{\partial \theta^2} + \frac{\partial}{\partial t}\left[(1-t^2)\frac{\partial S_m}{\partial t}\right] = 0, \tag{5.108}$$

which reduces to the Legendre equation for S_m as function of $t = \cos\varphi$ alone (this is no surprise because $H_m(x, y, z) = (r^m/r'^{\,m+1} 1)P_m(u)$, so that P_m is a surface harmonic of order m and a proper choice of axis allows $\varphi' = 0$, that gives $u = \cos\varphi$. andP_m does not depend on θ).

Equation (5.106) is invariant upon change of m with $-(m+1)$, implying that to every solid harmonic $r^m S_m$ corresponds the harmonic $r^{-(m+1)} S_m$.

5.11 Solution of the Differential Equation for $S_m(\theta, \varphi)$

We seek $2m+1$ particular solutions of the differential equation (5.108) by separation of variables. We denote them by $S_m^p(\theta, t)$, with[10]

$$S_m^p(\theta, t) = f(\theta)g(t), \quad p = 1, \ldots, 2m+1. \tag{5.109}$$

A linear combination of the $2m+1$ solutions $S_m^p\theta, t)$ will then represent the most general solution

[10] Note that here and in the following p and q are superscripts indices and not exponents.

$$S_m(\theta, t) = \sum_{p=1}^{2m+1} \alpha_p S_m^p(\theta, t), \tag{5.110}$$

where α_p are $2m+1$ constants.

Substituting $S_m(\theta, t)$ with $f(\theta)g(t)$ in Eq. 5.108 leads to

$$m(m+1)fg + \frac{1}{1-t^2} g \frac{\partial^2 f}{\partial \theta^2} + f \frac{\partial}{\partial t}\left[(1-t^2)\frac{\partial g}{\partial t}\right] = 0, \tag{5.111}$$

which can be written, separating in θ and t,[11]

$$m(m+1)(1-t^2) + \frac{1-t^2}{g}\frac{\mathrm{d}}{\mathrm{d}t}\left[(1-t^2)\frac{\mathrm{d}g}{\mathrm{d}t}\right] = -\frac{1}{f}\frac{\mathrm{d}^2 f}{\mathrm{d}\theta^2} \tag{5.112}$$

which is admissible only if both sides of this equation are equal to the same real constant, that we call ν, i.e.

$$m(m+1)(1-t^2) + \frac{1-t^2}{g}\frac{\mathrm{d}}{\mathrm{d}t}\left[(1-t^2)\frac{\mathrm{d}g}{\mathrm{d}t}\right] = \nu, \tag{5.113}$$

and

$$-\frac{1}{f}\frac{\mathrm{d}^2 f}{\mathrm{d}\theta^2} = \nu. \tag{5.114}$$

Because the characteristic equation associated to the differential equation (5.112) is $\xi^2 + \nu = 0$, a non divergent solution for S_m^p requires $\nu \geq 0$.[12] Additionally, to have a periodic solution of period 2π—in order to guarantee the univocity of S_m^p on the unit sphere—it must be $\sqrt{\nu} = q \in \mathbb{Z}^{\pm}$,[13] i.e. $\nu = q^2$. Finally

$$f(\theta) = a\cos(q\theta) + b\sin(q\theta), \qquad q = 0, \pm 1, \pm 2, \dots . \tag{5.115}$$

Once that a particular integral of Eq. 5.113, indicated by $P_m^q(t)$, is known, a particular solution of the original Eq. 5.108 is given by

$$P_m^q(t)f(\theta) = P_m^q \times \begin{cases} a\cos(q\theta) + b\sin(q\theta) \\ a\cos(-q\theta) + b\sin(-q\theta). \end{cases} \tag{5.116}$$

---

[11] It suffices multiplying Eq. 5.111 by $\dfrac{1-t^2}{fg}$.

[12] In such case the solution can be written $f(\theta) = a\cos(\sqrt{\nu}\theta) + b\sin(\sqrt{\nu}\theta)$.

[13] $\cos[\sqrt{\nu}(\theta + 2\pi)] = \cos(\sqrt{\nu}\theta)\cos(\sqrt{\nu}2\pi) - \sin(\sqrt{\nu}\theta)\sin(\sqrt{\nu}2\pi) = \cos\sqrt{\nu}\theta$ only if $\sqrt{\nu}$ is a positive or negative integer (similar reasoning for $\sin[\sqrt{\nu}(\theta + 2\pi)]$).

Then, any linear combination of the type

$$\begin{aligned} S_m^q &= \alpha P_m^q[a\cos(q\theta) + b\sin(q\theta]) + \beta P_m^q[a\cos(-q\theta) + b\sin(-q\theta)] = \\ &= (\alpha+\beta)P_m^q a\cos(q\theta) + (\alpha-\beta)P_m^q b\sin(q\theta) \equiv S_{m,1}^q + S_{m,2}^q, \end{aligned} \tag{5.117}$$

is still a solution of the equation of surface spherical harmonics.

A particular solution in terms of cosines only is obtained for $\alpha = \beta$, while it is in terms of sines only when $\alpha = -\beta$. The further constraints $\alpha = \beta = \frac{1}{2a}$ and $\alpha = -\beta = -\frac{1}{2b}$ lead to the so-called Laplace functions of order q

$$S_{m,1}^q = P_m^q \cos(q\theta), \qquad S_{m,2}^q = P_m^q \sin(q\theta). \tag{5.118}$$

Clearly, the most general solution of the surface spherical harmonic equation is given by summation of terms like the two in the rightmost side of Eq. 5.117 varying q between 0 and m (so that q assumes the $2m+1$ integer values $= 0, \pm1, \pm2, \ldots, \pm m$), as is the case for a homogeneous and harmonic polynomial of degree m. Defining $a_{mp} = (\alpha+\beta)a$ and $b_{mp} = (\alpha-\beta)b$, it results

$$\begin{aligned} S_m(\theta,\varphi) &= \sum_{p=0}^{m} S_m^p(\theta,\varphi) = \sum_{p=0}^{m} \left(a_{mp}P_m^p\cos(p\theta) + b_{mp}P_m^p\sin(p\theta)\right) = \\ &= a_{m0}P_m^0 + 0 + a_{m1}P_m^1\cos\theta + b_{m1}P_m^1\sin\theta + \cdots + \\ &+ a_{mm}P_m^m\cos(m\theta) + b_{mm}P_m^m\sin(m\theta) \qquad (2m+1 \text{ terms}) \,. \end{aligned} \tag{5.119}$$

Obviously, each single term in the above expression is also a surface spherical harmonic, individually obtained by choices of the kind $b_{mk} = 0,\ \forall k,\ a_{mk} = 0\,\forall k \neq p$, and $a_{mp} = 1$, leading to

$$S_m^p(\theta,\varphi) = P_m^p\cos(p\theta), \tag{5.120}$$

or $a_{mk} = 0,\ \forall k,\ b_{mk} = 0\,\forall k \neq p$, and $b_{mp} = 1$, leading to

$$S_m^p(\theta,\varphi) = P_m^p\sin(p\theta,) \tag{5.121}$$

called often *elementary* surface spherical harmonics.

5.12 The Associated Legendre functions $P_m^p(\cos\varphi)$

Equation 5.111 for the generic $\nu = p^2$, after division by $1-t^2 \neq 0$, reads as

$$m(m+1)g(t) + \frac{\mathrm{d}}{\mathrm{d}t}\left[(1-t^2)\frac{dg}{dt}\right] = \frac{p^2}{1-t^2}g(t), \tag{5.122}$$

that for $p = 0$ reduces to

$$m(m+1)g(t) + \frac{\mathrm{d}}{\mathrm{d}t}\left[(1-t^2)\frac{\mathrm{d}g}{\mathrm{d}t}\right] = 0, \tag{5.123}$$

that is Legendre differential equation, for which one particular solution is known to be[14]

$$P_m^0(t) = P_m(t) = P_m(\cos\varphi). \tag{5.124}$$

If $p \neq 0$, the differential equation satisfied by P_m^p is called the *associated* Legendre differential equation, whose solution in terms of Legendre polynomials is due to Poincarè

$$P_m^p(t) = (t^2-1)^{\frac{p}{2}}\frac{\mathrm{d}^p}{\mathrm{d}t^p}P_m(t), \qquad p \leq m; \tag{5.125}$$

and, using Rodrigues' formula

$$P_m^p(t) = \frac{1}{2^m m!}(t^2-1)^{\frac{p}{2}}\frac{\mathrm{d}^{p+m}}{\mathrm{d}t^{p+m}}(t^2-1)^m\ . \tag{5.126}$$

The $P_m^p(t)$ functions are called *associated* Legendre functions of the *first kind*.
Let us verify that the expression 5.126 actually satisfies the Eq. 5.122.
By introducing the auxiliary function $z(t)$ defined as

$$g(t) = (t^2-1)^{p/2}z(t) \tag{5.127}$$

the Eq. 5.122 becomes

$$(1-t^2)z'' - 2t(p+1)z' + (m^2+m-p^2-p)z = 0 \tag{5.128}$$

where the prime stays for derivative with respect to t. On the other side, a p-times differentiation of the Legendre equation gives

$$m(m+1)\frac{\mathrm{d}^p}{\mathrm{d}t^p}P_m + \frac{\mathrm{d}^{p+1}}{\mathrm{d}t^{p+1}}\left[(1-t^2)P_m'\right] = 0. \tag{5.129}$$

By using the following expansion[15]

$$\frac{\mathrm{d}^{p+1}}{\mathrm{d}t^{p+1}}(1-t^2)P_m' = \sum_{k=0}^{p+1}\binom{p+1}{k}\frac{\mathrm{d}^k}{\mathrm{d}t^k}(1-t^2)\frac{\mathrm{d}^{p+1-k}}{\mathrm{d}t^{p+1-k}}P_m' =$$

[14] So $P_m(t)$ is the Laplace function of order zero.

[15] The symbol $[p]$ stands for pth derivative with respect to t.

$$= \sum_{k=0}^{p+1} \binom{p+1}{k} (1-t^2)^{[k]} P_m^{[p+2]} = \tag{5.130}$$

$$= (1-t^2) P_m^{[p+2]} + (p+1)(-2t) P_m^{[p+1]}$$

$$+ \frac{(p+1)!}{2!(p-1!)} (-2) P_m^{[p]},$$

and making a substitution in Eq. 5.129, we arrive to

$$m(m+1) P_m^{[p]} + (1-t^2) P_m^{[p+2]} - 2t(p+1) P_m^{[p+1]} - p(p+1) P_m^{[p]} =$$

$$= (1-t^2) P_m^{[p+2]} - 2t(p+1) P_m^{[p+1]} + (m^2 + m - p^2 - p) P_m^{[p]} = 0, \tag{5.131}$$

where the last equality corresponds to satisfying Eq. 5.128 with $z = P_m^{[p]}$, so that the solution $g(t)$ of Eq. 5.122 is indeed

$$g(t) = P_m^p(t) = (t^2 - 1)^{\frac{p}{2}} P_m^{[p]}(t) \, . \tag{5.132}$$

It is also possible to verify that the set of functions $P_m^p(t)$ above defined is an orthogonal system in $(-1, 1)$, giving

$$(P_m^p, P_n^p) = \int_{-1}^{1} P_m^p P_n^p \mathrm{d}t = (-1)^p \frac{(m+p)!}{(m-p)!} \frac{2}{2m+1} \delta_{mn},$$

$$\forall p \geq 0, \;\; m = p, p+1, p+2, \ldots; \;\; n = p, p+1, p+2, \ldots \tag{5.133}$$

5.13 Note on the Associated Legendre Equation

The associated Legendre differential equation with m and p non negative integers has two families of solutions, so that its general solution is

$$g(t) = c_1 P_m^p + c_2 Q_m^p, \tag{5.134}$$

where Q_m^p are called associated Legendre functions of the *second kind*

$$Q_m^p = (-1)^p (t^2 - 1)^{\frac{p}{3}} \frac{\mathrm{d}^p Q_m}{\mathrm{d}t^p} \, , \tag{5.135}$$

where $Q_m(t)$ are the Legendre functions of the second kind, forming the second linearly independent solution to Legendre equation, alongside $P_m(t)$. Being t real and $|t| \leq 1$, the function $P_m^p(t)$ in Eq. 5.123 is real only for even values of p, so that it is often advantageous to replace it with

$$\overline{P}_m^p(t) = (1-t^2)^{\frac{p}{2}} \frac{\mathrm{d}^p}{\mathrm{d}t^p} P_m(t), \tag{5.136}$$

and/or with

$$T_m^p(t) = (-1)^p (1-t^2)^{\frac{p}{2}} \frac{\mathrm{d}^p}{\mathrm{d}t^p} P_m(t), \tag{5.137}$$

these latter real functions being called *Ferrers functions* of the first kind (we will use them in the following, indicating them as P_m^p).

Exercise 5.4 Determine the conditions on the function $f(x, y, z)$, assumed to depend on x, y, z only through the linear combination $u = ax + by + cz$ (with real constants a, b, c), for f to be harmonic.

Solution Given $u = ax + by + cz$, it is

$$\nabla^2 f = \frac{\partial^2 f}{\partial x^2} + \frac{\partial^2 f}{\partial y^2} + \frac{\partial^2 f}{\partial z^2} = a^2 \frac{\mathrm{d}^2 f}{\mathrm{d}u^2} + b^2 \frac{\mathrm{d}^2 f}{\mathrm{d}u^2} + c^2 \frac{\mathrm{d}^2 f}{\mathrm{d}u^2} = (a^2 + b^2 + c^2) \frac{\mathrm{d}^2 f}{\mathrm{d}u^2},$$

which is zero either by the trivial condition $a^2 + b^2 + c^2 = 0$ or $\dfrac{\mathrm{d}^2 f}{\mathrm{d}u^2} = 0$, this latter meaning that f is linear in u, that is

$$f(x, y, z) = \alpha u + \beta = \alpha(ax + by + cz) + \beta\ .$$

Exercise 5.5 Given a harmonic function $F(x, y, z)$, determine under what conditions F/r is also harmonic.

Solution The condition is $\nabla^2(F/r) = 0$, where

$$\begin{aligned}
\nabla^2 \frac{F}{r} &= \nabla \cdot \nabla \frac{F}{r} = \nabla \cdot \left(\frac{1}{r} \nabla F - \frac{F}{r^2} \nabla r \right) = \\
&= \frac{1}{r} \nabla^2 F + \nabla F \cdot \nabla \frac{1}{r} - \frac{F}{r^2} \nabla^2 r + \nabla r \cdot \nabla \left(-\frac{F}{r^2} \right) = \\
&= 0 - \frac{1}{r^2} \nabla F \cdot \mathbf{e}_r - \frac{F}{r^2} \frac{2}{r} - \mathbf{e}_r \nabla \frac{F}{r^2} = \\
&= -\frac{1}{r^2} \frac{\partial F}{\partial r} - \frac{2}{r^3} F - \mathbf{e}_r \cdot \left(-\frac{2}{r^3} F \mathbf{e}_r + \frac{1}{r^2} \nabla F \right) = \\
&= -\frac{2}{r^2} \frac{\partial F}{\partial r}.
\end{aligned}$$

5.14 Zonal, Tesseral and Sectorial Spherical Harmonics

The $2m + 1$ elementary surface spherical harmonics of order m can be written, for chosen set of constants a_{mp}, b_{mp}, as

- $a_{m0} P_m(t)$, for $p = 0$ (1 harmonic),
  and
- $P_m^p(t) \left[a_{mp} \cos(p\theta) + b_{mp} \sin(p\theta)\right]$, for $1 \le p \le m$ (2m harmonics).

We can distinguish three cases that characterize the locations on the unitary sphere of the zeros of these harmonics

(*i*) $p = 0$: *zonal harmonics,*
(*ii*) $1 \le p < m$: *tesseral harmonics,*
(*iii*) $p = m$: *sectorial harmonics.*

(*i*) case $p = 0$: **zonal harmonics**

If $p = 0$, then the surface spherical harmonic $P_m^0(t)$ coincides with the Legendre polynomial in the variable $t = \cos\varphi$, This polynomial has m distinct roots, symmetrical distributed with respect to $t = \cos\varphi = 0$, which corresponds to $\varphi\pi/2$. When m is odd, $t = 0$ (i.e. $\varphi = \pi/2$) is itself one of the roots (see Fig. 5.5). The roots lie, therefore, on *parallel* circles which divide the unitary sphere in zones, hence the name of *zonal* harmonics. Recall that at the poles ($\varphi = 0$, $\varphi = \pi$) $P_m(1) = 1$ and $P_m(-1) = (-1)^m$.

(*ii*) case $1 \le p \le m$: **tesseral harmonics**

If $1 \le p \le m$, the harmonic takes the Ferrers form

$$(-1)^p \left[(1-t^2)^{\frac{p}{2}} \frac{1}{2^m m!} \frac{\mathrm{d}^{m+p}}{\mathrm{d}t^{m+p}} (t^2-1)^m\right] \left[a_{mp} \cos(p\theta) + b_{mp} \sin(p\theta)\right], \tag{5.138}$$

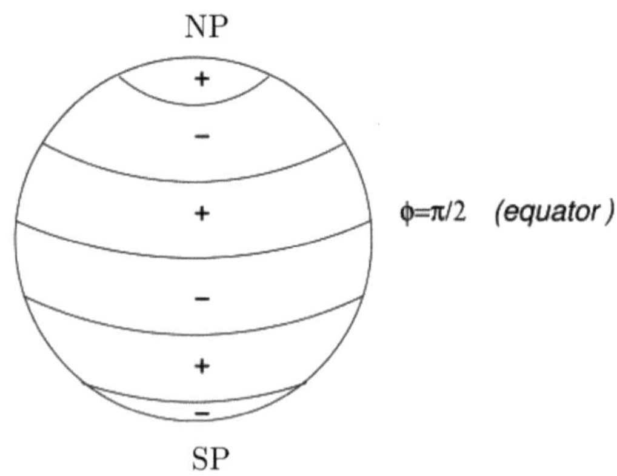

Fig. 5.5 Unitary sphere zone subdivision for $P_5(\varphi)$. Symbol + or − identify the six regions of positivity or negativity of the polynomial P_5. NP and SP refer to North ($\varphi = 0$) and South ($\varphi = \pi$) poles

whose roots are those of

(a) $(1-t^2)^{\frac{p}{2}}\dfrac{1}{2^m m!}\dfrac{\mathrm{d}^{m+p}}{\mathrm{d}t^{m+p}}(t^2-1)^m=0,$
and/or
(b) $a_{mp}\cos(p\theta)+b_{mp}\sin(p\theta)=0.$

Case (a) corresponds to

$$t^2-1=0 \Longleftrightarrow \cos\varphi=\pm 1 \Longleftrightarrow \varphi=0,\pi \quad \text{(North and South poles)}, \tag{5.139}$$

and/or

$$\frac{1}{2^m m!}\frac{\mathrm{d}^{m+p}}{\mathrm{d}t^{m+p}}(t^2-1)^m=\frac{\mathrm{d}^p}{\mathrm{d}t^p}P_m(t)=0, \tag{5.140}$$

this latter equation (as it could be demonstrated) having $m-p$ real roots in the interval $-1<t<1$ symmetrically distributed as for the zonal harmonics.

Case (b) corresponds to

$$\tan(p\theta)=-\frac{a_{mp}}{b_{mp}}, \tag{5.141}$$

so that

$$\theta_k=\frac{1}{p}\arctan\left(-\frac{a_{mp}}{b_{mp}}\right)+\frac{k}{p}\pi, \quad k=0,1,2,\cdots,p-1. \tag{5.142}$$

The values of θ_k above define p great circles on the unit sphere, spaced at intervals of $\Delta\theta=\theta_k-\theta_{k-1}=\pi/p$. The $2m-2$ zeros lie at the poles and on both parallel and (meridian) great circles which divide the sphere surface into a set of tesseras of alternating sign for the harmonic, from which the name of *tesseral* harmonics (Fig. 5.6).

The sign of $S_m^p(\theta,\varphi)$ changes crossing a parallel and a meridian.

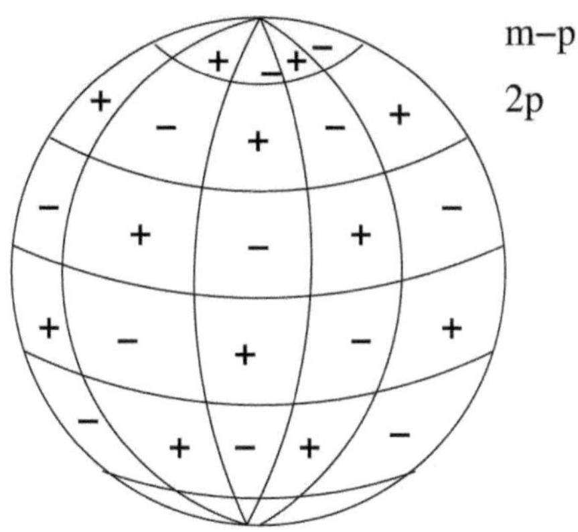

Fig. 5.6 Sign distribution of a spherical harmonic $S_m^p(\theta,\varphi)$. There are $m-p$ parallels and $2p$ meridians

(*iii*) case $p = m$: **sectorial harmonics** If $p = m$ the generic harmonic is

$$(-1)^m \left[(t^2-1)^{\frac{m}{2}} \frac{1}{2^m m!} \frac{\mathrm{d}^{2m}}{\mathrm{d}t^{2m}} (t^2-1)^m \right] [a_{mm}\cos(m\theta) + b_{mm}\sin(m\theta)] . \tag{5.143}$$

It is easily seen that

$$\frac{\mathrm{d}^{2m}}{\mathrm{d}t^{2m}} (t^2-1)^m = (2m)!, \tag{5.144}$$

and so the zeros of the $P_m^m(\theta, \varphi)$ harmonic are obtained by

$$(t^2-1)^{\frac{m}{2}} = 0, \tag{5.145}$$

and/or

$$a_{mm}\cos(m\theta) + b_{mm}\sin(m\theta) = 0, \tag{5.146}$$

that is

$$\begin{cases} t = \pm 1 \Longleftrightarrow \varphi = 0, \pi, & \text{2 poles} \\ \theta_k = \arctan\left(-\dfrac{a_{mm}}{b_{mm}}\right) + \dfrac{k}{m}\pi, & m \text{ great circles.} \end{cases} \tag{5.147}$$

The zeros location on the unitary sphere along meridians and the subsequent subdivision in spherical sectors justifies the name of sectorial harmonics (Fig.5.7).

Exercise 5.6 Write the explicit expression of $P_2^1(t)$, $P_3^2(t)$, $P_2^3(t)$.

Fig. 5.7 Sign subdivision of a generic sectorial harmonic $S_m^m(\theta, \varphi)$

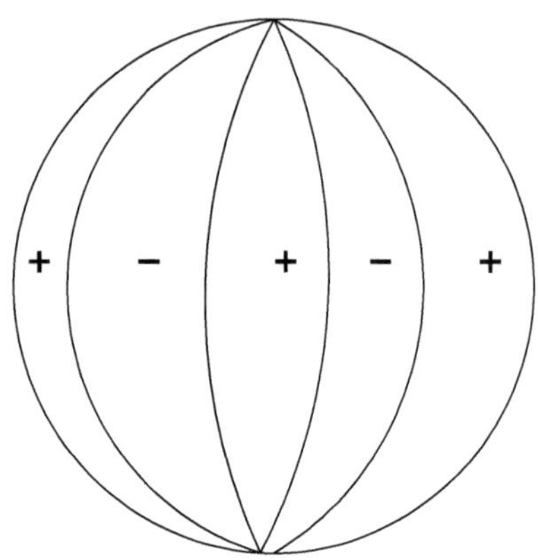

Solution Being

$$P_m^p(t) = (-1)^p(1-t^2)^{\frac{p}{2}} \frac{d^p}{dt^p} P_m(t),$$

it results

$$P_2^1(t) = -(1-t^2)^{\frac{1}{2}} \frac{d}{dt} P_2(t) = -(1-t^2)^{\frac{1}{2}} \frac{d}{dt}\left[\frac{1}{2}\left(3t^2-1\right)\right] = -3t\sqrt{1-t^2}$$

$$P_3^2(t) = (1-t^2)\frac{d^2}{dt^2} P_3(t) = (1-t^2)\frac{d^2}{dt^2}\left[\frac{1}{2}(5t^3-3t)\right] = 15t(1-t^2)$$

$$P_2^3(t) = -(1-t^2)^{\frac{3}{2}} \frac{d^3}{dt^3} P_2(t) = -(1-t^2)^{\frac{3}{2}} \frac{d^3}{dt^3}\left[\frac{1}{2}(3t^2-1)\right] = 0.$$

Exercise 5.7 Verify that $P_3^2(t)$ given in the Exercise 5.6 satisfies its associated Legendre equation.

Solution The associated Legendre equation (see Eq. 5.122)

$$m(m+1)g + \frac{d}{dt}\left[(1-t^2)g'\right] - \frac{p^2}{1-t^2}g = 0,$$

letting $p = 2$ and $m = 3$, becomes

$$12g + \frac{d}{dt}\left[(1-t^2)g'\right] - \frac{4}{1-t^2}g = 0.$$

Taking $g(t) = P_3^2(t)$ and substituting into the above equation

$$\begin{aligned} 180t(1-t^2) + \frac{d}{dt}\left[(1-t^2)(15-45t^2)\right] - \frac{4}{1-t^2}15t(1-t^2) &= \\ = 15t\left[12(1-t^2) + 12t^2 - 12\right] &= 0, \end{aligned}$$

as we wanted to show.

Exercise 5.8 Write the associated Legendre equation in the form $f(g, g', g''; t) = a(t)g'' + b(t)g' + c(t)g = 0$.

Solution The associated Legendre equation is

$$m(m+1)g - 2tg' + (1-t^2)g'' - \frac{p^2}{1-t^2}g = 0,$$

or, equivalently

$$(1-t^2)g'' - 2tg' + \left[m(m+1) - \frac{p^2}{1-t^2}\right] g = 0,$$

so that $a(t) = 1 - t^2$, $b(t) = -2t$, $c(t) = m(m+1) - p^2/(1-t^2)$.

Exercise 5.9 Show that for $0 \leq \varphi \leq \pi$

$$-\sin\varphi\cos(3\theta) = 8\left[7 \times \frac{1}{6!} P_3^3(t) + 11 \times \frac{2!}{8!} P_5^3(t) + 15 \times \frac{4!}{10!} P_7^3(t) + \cdots\right] \cos(3\theta).$$

Solution Left to the reader.

5.15 Orthogonality of Surface Spherical Harmonics

Theorem 5.7 *The set of elementary harmonics*

$$\{Y_m^p(\theta,\varphi)\} = \{P_m^p(\varphi)\cos(p\theta),\ P_m^p(\varphi)\sin(p\theta)\}, \qquad p = 0, 1, \cdots, m,$$

is a system of orthogonal functions on the unitary sphere.

Proof Compute (Y_m^p, Y_n^p), treating the various possible cases.

$$(a) \quad \int_{S_1(0)} P_m^p \cos(p\theta) P_n^p \sin(p\theta)\, \mathrm{d}\sigma = \int_0^{2\pi} \cos(p\theta)\sin(p\theta)\, \mathrm{d}\theta \times \int_0^{\pi} P_m^p P_n^p \sin\varphi\, \mathrm{d}\varphi = 0,$$

because

$$\int_0^{2\pi} \cos(p\theta)\sin(p\theta)\, \mathrm{d}\theta = 0.$$

$$(b) \quad \int_{S_1(0)} P_m^p \cos(p\theta) P_n^p \cos(p\theta)\, \mathrm{d}\sigma = \int_0^{2\pi} \cos^2(p\theta)\, \mathrm{d}\theta \int_0^{\pi} P_m^p P_n^p \sin\varphi\, \mathrm{d}\varphi =$$

$$= \pi(1+\delta_{0p}) \int_{-1}^{1} P_m^p P_n^p\, \mathrm{d}t = \pi(1+\delta_{0p})(P_m^p, P_n^p),$$

where we made use of that

$$\int_0^{2\pi} \cos^2(p\theta)\,\mathrm{d}\theta = \begin{cases} 2\pi, & \text{if } p = 0, \\ \frac{1}{p}\left[\dfrac{\sin(p\theta)\cos(p\theta)}{2} + \dfrac{p\theta}{2}\right]_{\theta=0}^{\theta=2\pi} = \pi, & \text{if } p \neq 0. \end{cases}$$

$$(c) \quad \int_{S_1(0)} P_m^p \sin(p\theta) P_n^p \sin(p\theta)\,\mathrm{d}\sigma = \int_0^{2\pi} \sin^2(p\theta)\,\mathrm{d}\theta \int_{-1}^{1} P_m^p P_n^p\,\mathrm{d}t = \\ = \pi(1-\delta_{0p})(P_m^p, P_n^p),$$

because

$$\int_0^{2\pi} \sin^2(p\theta)\,\mathrm{d}\theta = \begin{cases} 0, & \text{if } p = 0, \\ \frac{1}{p}\left[\dfrac{\sin(p\theta)\cos(p\theta)}{2} + \dfrac{p\theta}{2}\right]_{\theta=0}^{\theta=2\pi} = \pi, & \text{if } p \neq 0. \end{cases}$$

It remains to compute (letting $t = \cos\varphi$)

$$(P_m^p, P_n^p) = \int_{-1}^{1} P_m^p(t) P_n^p(t)\,\mathrm{d}t = (-1)^{2p} \int_{-1}^{1} (1-t^2)^p P_m^{[p]} P_n^{[p]}\,\mathrm{d}t,$$

which, integrating by parts[16] gives the following

$$(P_m^p, P_n^) = \left[\left(1-t^2\right) P_m^{[p]} P_n^{[p-1]}\right]_{-1}^{1} - \int_{-1}^{1} P_n^{[p-1]} \frac{\mathrm{d}}{\mathrm{d}t}\left[(1-t^2)^p P_m^{[p]}\right]\,\mathrm{d}t.$$

By means of Eq. 5.128 we have, after multiplication by $(1-t^2)^p$

$$\frac{\mathrm{d}}{\mathrm{d}t}\left[(1-t^2)^{p+1} P_m^{[p+1]}\right] = (p^2 + p - m^2 - m)(1-t^2)^p P_m^{[p]} = \\ = -(m-p)(p+m+1)(1-t^2)^p P_m^{[p]},$$

that, by substitution in the above integral (provided the change $p+1 \longrightarrow p$) gives

$$\left(P_m^p, P_n^p\right) = (m-p+1)(m+p) \int_{-1}^{1} P_m^{[p-1]} P_m^{[p-1]} (1-t^2)^{p-1}\,\mathrm{d}t =$$

---

[16] With $u = \left(1-t^2\right)^p P_m^{[p]}$ and $dv = P_m^{[p]} dt$; remember that the symbol $[p]$ here means pth-derivative with respect to t.

$$= (m-p+1)(m+p)\left(P_m^{p-1}P_n^{p-1}\right) =$$

$$= (m-p+1)(m+p)(m-p+1)(p-1+m)\left(P_m^{p-2}P_n^{p-2}\right) =$$

$$= \cdots = \frac{(m+p)!}{(m-p)!}\frac{2}{2m+1}\delta_{mn}.$$

In conclusion, we have shown that

$$\left(Y_m^p, Y_n^p\right) = (1 \pm \delta_{0p})\pi \frac{(m+p)!}{(m-p)!}\frac{2}{2m+1}\delta_{mn},$$

that proves the theorem.

Thanks to the above result, the set of $\left\{Y_m^p\right\}$ for every $p = 0, 1, \cdots, m$, results to be a system of orthogonal functions for $0 \le \theta < 2\pi$ and $0 \le \varphi \le \pi$, that can be used as basis functions for expansion in series of functions satisfying the Dirichlet conditions (see Sect. 5.7).

By the result of this theorem, it is easily shown (left to the reader as exercise) that, if $S_m(\theta, \varphi)$ and $\overline{S}_n(\theta, \varphi)$ are surface spherical harmonics, then

$$\begin{cases} (S_m, \overline{S}_n) = 0, & \text{if } m \ne n, \\ (S_m, \overline{S}_m) = \dfrac{2\pi}{2m+1}\left[2a_{m0}\overline{a}_{m0} + \sum_{k=1}^m (a_{mk}\overline{a}_{mk} + b_{mk}\overline{b}_{mk})\dfrac{(m+k)!}{(m-k)!}\right], \end{cases} \tag{5.148}$$

where

$$\begin{aligned} S_m(\theta, \varphi) &= \sum_{p=0}^{n} (a_{np}\cos(p\theta) + b_{np}\sin(p\theta))\, P_m^p, \\ \overline{S}_n(\theta, \varphi) &= \sum_{p=0}^{n} (\overline{a}_{np}\cos(p\theta) + \overline{b}_{np}\sin(p\theta))\, P_n^p. \end{aligned} \tag{5.149}$$

In particular, if $\overline{S}_m = P_m(\cos\varphi)$ (as from the choices $\overline{a}_{m0} = 1$ and $\overline{a}_{mp} = \overline{b}_{mp}, \quad \forall p > 0$), it results

$$(S_m, P_m) = \frac{4\pi}{2m+1}a_{m0} = \frac{4\pi}{2m+1}S_m(0,0), \tag{5.150}$$

due to that

$$S_m(0,0) = a_{m0}P_m(1) + \sum_{p=1}^{m} a_{mp}P_m^p(1) = a_{m0}, \tag{5.151}$$

being $P_m(1) = 1 \quad \forall m$, and $P_m^p(1) = 0 \quad \forall m$, with $p \le m$.

5.16 Expansions in Terms of Spherical Harmonics

Given that the set of $Y_m^p(\theta,\varphi)$ constitutes a basis, a sufficiently regular function $f(\theta,\varphi)$ can be expanded as

$$f(\theta,\varphi)=\sum_{m=0}^{\infty}\sum_{p=0}^{m}(a_{mp}\cos(p\theta)+b_{mp}\sin(p\theta))P_m^p(\cos\varphi), \tag{5.152}$$

whose coefficients a_{mp} (for $p=0,1,2,\cdots,m$) and b_{mp} (for $p=1,2,\cdots,m$), are

$$a_{mp}=\frac{\int\limits_{S_1(0)} f\,P_m^p\cos(p\theta)\,d\sigma}{\int\limits_{S_1(0)}(P_m^p\cos(p\theta))^2\,d\sigma}=\frac{2m+1}{2\pi(1+\delta_{p0})}\frac{(m-p)!}{(m+p)!}\int\limits_{S_1(0)} f\,P_m^p\cos(p\theta)\,d\sigma,$$

$$b_{mp}=\frac{\int\limits_{S_1(0)} f\,P_m^p\sin(p\theta)\,d\sigma}{\int\limits_{S_1(0)}(P_m^p\sin(p\theta))^2\,d\sigma}=\frac{2m+1}{2\pi}\frac{(m-p)!}{(m+p)!}\int\limits_{S_1(0)} f\,P_m^p\sin(p\theta)\,d\sigma, \tag{5.153}$$

as obtained by multiplication of both sides of Eq. 5.152 by $P_m^p\cos(p\theta)$ and $P_m^p\sin(p\theta)$ and integration over the unitary sphere, keeping in mind what obtained in the preceding Sect. 5.15.

Note that the coefficient b_{m0} is not defined because it does not contribute to the expansion (5.152) (it multiplies $\sin 0=0$).

Note 16
The unique determination of coefficients of the series means that if the above series expansion of $f(\theta,\varphi)$ exists it is unique.

5.17 Expansion of the Potential Generated by a Regular Mass Distribution

Remembering that

$$\frac{1}{\|\mathbf{r}-\mathbf{r}'\|}=\frac{1}{r'}\sum_{m=0}^{\infty}P_m(\cos\gamma)\mu^m=\sum_{m=0}^{\infty}P_m(\cos\gamma)\frac{r^m}{r'^{\,m+1}} \tag{5.154}$$

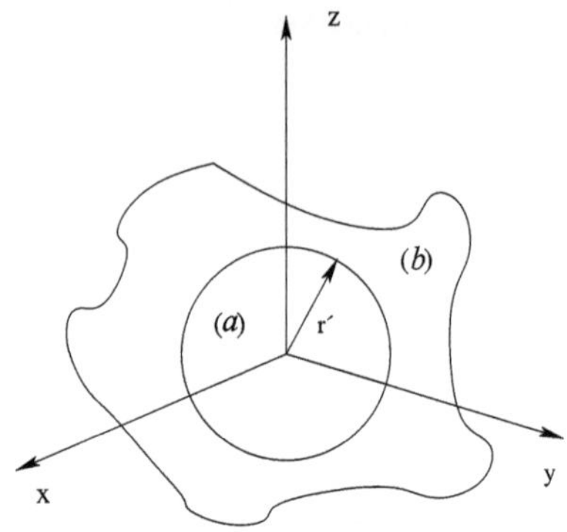

Fig. 5.8 Sketch of the matter domain C, related to the potential expansion

is convergent for $\mu = r/r' < 1$, it is clear that if we divide the mass domain (where $\rho(\mathbf{r}) \geq 0$) in the two parts (a) $r/r' < 1$, and (b) $r/r' > 1$,
the following two expansions hold, respectively

$$(a) \qquad \frac{1}{\|\mathbf{r}-\mathbf{r}'\|} = \sum_{m=0}^{\infty} P_m(\cos\gamma)\,\frac{r^m}{{r'}^{m+1}}, \quad \text{convergent for } \frac{r}{r'} < 1,$$
$$(b) \qquad \frac{1}{\|\mathbf{r}-\mathbf{r}'\|} = \sum_{m=0}^{\infty} P_m(\cos\gamma)\,\frac{{r'}^m}{r^{m+1}}, \quad \text{convergent for } \frac{r'}{r} < 1. \tag{5.155}$$

Example 5.17.1 Let $C = \{\mathbf{r} : \rho(\mathbf{r}) \geq 0\}$ be the domain of mass M and consider a generic point $\mathbf{r}' \in C$. Let also consider the sphere of radius r' centred at the origin of the reference system.

In Fig. 5.8 the region (a) represents the open ball $S_{r'}(0)$, and the region (b) represents $C \setminus \overline{S}_{r'}(0)$, where $\overline{S}_{r'}(0)$ indicates the closed ball of radius r' and centre in the origin.

The two series expansions above can be considered as *internal* and *external* to $\overline{S}_{r'}(0)$, i.e. due to the contribution to the Newtonian potential $U(\mathbf{r})$ given by the matter inside and out the ball.

The potential in the generic point $\mathbf{r}' = (x', y', z')$, in its integral form is

$$U(\mathbf{r}') = G\int_C \frac{\mathrm{d}m}{\|\mathbf{r}-\mathbf{r}'\|} = G\Bigg[\sum_{m=0}^{\infty}\iiint_{S_{\mathbf{r}'}(0)} P_m(\cos\gamma)\frac{r^m}{{r'}^{m+1}1}\,\rho(\mathbf{r})\,\mathrm{d}^3\mathbf{r} +$$
$$+\sum_{m=0}^{\infty}\iiint_{\mathbb{R}^3\setminus S_{r'}(0)} P_m(\cos\gamma)\frac{{r'}^m}{r^{m+1}}\,\rho(\mathbf{r})\,\mathrm{d}^3\mathbf{r}\Bigg] = U_{int}(\mathbf{r}') + U_{ext}(\mathbf{r}'). \tag{5.156}$$

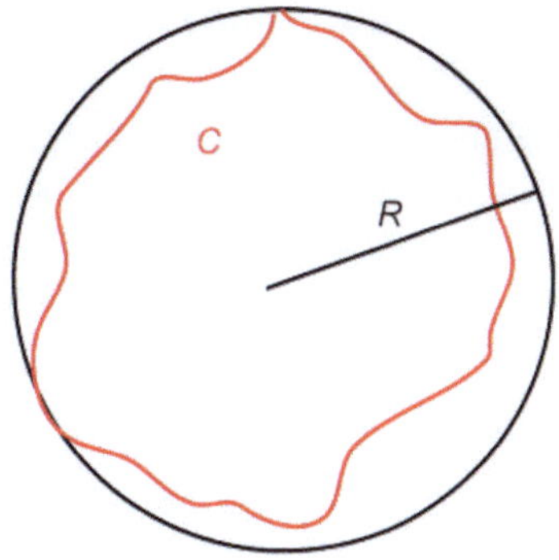

Fig. 5.9 The radius R is that of the smallest sphere containing the generic mass domain C, outlined in red

The integration term by term is possible because the two series above are uniformly convergent because bounded by the geometric series of argument $\mu < 1$,[17]. The resulting expansion is

$$U(\mathbf{r}') = \sum_{m=0}^{\infty} U_{int,m} \frac{1}{{r'}^{m+1}1} + \sum_{m=0}^{\infty} U_{ext,m} {r'}^{m}, \tag{5.157}$$

where the $U_{int,m}$ and $U_{ext,m}$ coefficients depend also on the direction of vector radius $\mathbf{r}'$, and are given by

$$U_{int,m} = G \iiint_{S_{\mathbf{r}'}(0)} P_m(\cos\gamma) r^m \rho(\mathbf{r})\,\mathrm{d}^3\mathbf{r}, \tag{5.158}$$

$$U_{ext,m} = G \iiint_{\mathbb{R}^3 \setminus S_{\mathbf{r}'}(0)} P_m(\cos\gamma) \frac{1}{r^{m+1}} \rho(\mathbf{r})\,\mathrm{d}^3\mathbf{r}. \tag{5.159}$$

Obviously, to the evaluation of the potential in points out of the minimum sphere containing C, whose radius we indicate with R (see Fig. 5.9), the contribution comes only from the *internal* terms. This means that an evaluation of the potential in the space between the surface delimiting C and the surface of the ball of radius R requires a direct evaluation of the potential integral, because here the series expansion is not valid.

[17] Note that $|P_m(\cos\gamma)| \leq 1$.

5.17.1 External Potential Series Expansion

We now deal with the potential series expansion in the external convergence region, $r' > R$, that is

$$U(\mathbf{r}') = U_{int}(\mathbf{r}') = G\sum_{m=0}^{\infty}\int\!\!\int\!\!\int_{S_R(0)} P_m(\cos\gamma)\frac{r^m}{r'^{\,m+1}1}\rho(\mathbf{r})\,\mathrm{d}^3\mathbf{r} = \\ = \sum_{m=0}^{\infty} U_m(\theta',\varphi')\frac{1}{r'^{\,m+1}}, \tag{5.160}$$

where we abbreviated $U_{int,m}$ in U_m.

The problem is, now, the evaluation of the integrals

$$U_m(\theta',\varphi') = G\int\!\!\int\!\!\int_{S_R(0)} P_m(\cos\gamma) r^m \rho(\mathbf{r})\,\mathrm{d}^3\mathbf{r}, \tag{5.161}$$

which are called *moments* of the matter distribution (they are indeed the average of the generic mth power of r weighted by the function $G\,P_m(\cos\gamma)\,\rho(\mathbf{r})$). In spherical polar coordinates the above integral takes the form

$$U_m(\theta',\varphi') = G\int\!\!\int\!\!\int_{S_R(0)} P_m(\cos\gamma) r^{m+2}\rho(\mathbf{r})\sin\varphi\,\mathrm{d}r\,\mathrm{d}\theta\,\mathrm{d}\varphi, \tag{5.162}$$

with $\cos\gamma = \cos\varphi\cos\varphi' + \sin\varphi\sin\varphi'\cos(\theta-\theta')$.

Let us now compute the two first coefficients, U_0 e U_1.

- If $m = 0$, then

$$P_0(\cos\gamma) = 1 \quad\Longrightarrow\quad U_0 = G\iiint_C \rho(\mathbf{r})\,\mathrm{d}^3\mathbf{r} = GM, \tag{5.163}$$

hence the 1st term in the series 5.160 is

$$U_0\frac{1}{r'} = \frac{GM}{r'}, \tag{5.164}$$

which is the *monopole* term. It corresponds to the lowest order approximation of the potential as that due by a point of mass M in the coordinate origin, approximation which is more and more valid at larger distances from the origin because the contribution of higher moments reduces as $1/r'^{\,m+1}1$ at large distances whilst the coefficients U_m are bound.

- If $m = 1$, then

$$P_1(\cos\gamma) = \cos\gamma = \frac{\mathbf{r}\cdot\mathbf{r}'}{rr'} \quad \Longrightarrow U_1 = G\iiint_C \frac{\mathbf{r}\cdot\mathbf{r}'}{rr'} r\rho(\mathbf{r})\,\mathrm{d}^3\mathbf{r} =$$
$$= G\frac{\mathbf{r}'}{r'}\cdot\int_C \mathbf{r}\,\mathrm{d}m = \frac{GM}{r'}\mathbf{r}'\cdot\mathbf{r}_G, \tag{5.165}$$

where $\mathbf{r}_G = \dfrac{\int_C \mathbf{r}\,\mathrm{d}m}{\int_C \mathrm{d}m}$ gives the centre of mass position of the matter domain C.

The term $U_1/r'^{\,2}$ in the series is the *dipole* term. It vanishes upon the choice of the coordinate origin in the mass distribution centre of mass (i.e. letting $\mathbf{r}_G = 0$). With other coordinate choices, the dipole term is equal to the component of the vector $GM\mathbf{r}'$ along $\mathbf{r}_G$.

The term with $m = 2$ is called *quadrupole* term, and so on. We limit to note that, upon the choice of z-axis in the direction of $\mathbf{r}'$, the quadrupole coefficient is

$$U_2 = \frac{1}{2}G\int_0^{\pi}\int_0^{2\pi}\int_0^{R} (3\cos\varphi - 1)\,r^4\rho(r,\theta,\varphi)\sin\varphi\,\mathrm{d}r\,\mathrm{d}\theta\,\mathrm{d}\varphi. \tag{5.166}$$

Now we look for a valid way to express the generic $U_m(\theta',\varphi')$.

At this scope, we recall that $P_m(\cos\gamma)$ is a surface spherical harmonic in θ and φ (thinking of the pair (θ,φ) as variables upon the unit sphere and of (θ',φ') as fixed[18]) so that it can be written on the harmonic basis

$$P_m(\cos\gamma) = a_{m0}P_m(\cos\varphi) + \sum_{p=1}^{m}\left[a_{mp}\cos(p\theta) + b_{mp}\sin(p\theta)\right]P_m^p(\cos\varphi), \tag{5.167}$$

where coefficients a_{mp} and b_{mp} depend upon the angular coordinates (θ',φ') of $\mathbf{r}'$ on the unit sphere.

The method to find a_{mp} and b_{mp} is the same as for the series expansion of the generic function $f(\theta,\varphi)$ (see 5.152), leading to

$$a_{m0} = \frac{2m+1}{4\pi}\int_C P_m(\cos\gamma)P_m(\cos\varphi)\,\mathrm{d}\sigma,$$

$$a_{mp} = \frac{2m+1}{2\pi}\frac{(m-p)!}{(m+p)!}\int_C P_m(\cos\gamma)P_m^p(\cos\varphi)\cos(p\theta)\,\mathrm{d}\sigma \quad (p = 1, 2, \cdots, m),$$

[18] Or vice versa.

$$b_{mp} = \frac{2m+1}{2\pi}\frac{(m-p)!}{(m+p)!}\int\limits_C P_m(\cos\gamma)P_m^p(\cos\varphi)\sin(p\theta)\,\mathrm{d}\sigma \quad (p=1,2,\cdots,m). \tag{5.168}$$

The calculation of the above integrals takes advantage by a proper change of reference frame, centring at (θ', φ'), letting $\overline{\theta} = \theta - \theta'$ and $\overline{\varphi} = \varphi - \varphi'$.

Upon this transformation $P_m(\cos\gamma) = P_m(\cos\overline{\varphi})$ so that by relation 5.150 it results

$$\begin{aligned}
(X_m, Y_m) = a_{m0} &= \frac{2m+1}{4\pi}\int\limits_C P_m(\cos\gamma)P_m(\cos\varphi)\,\mathrm{d}\sigma = P_m(\cos\varphi'),\\
a_{mp} &= 2\frac{(m-p)!}{(m+p)!}P_m^p(\cos\varphi')\cos(p\theta'),\\
b_{mp} &= 2\frac{(m-p)!}{(m+p)!}P_m^p(\cos\varphi')\sin(p\theta').
\end{aligned} \tag{5.169}$$

Due to the above result the series coefficients are simply

$$\begin{aligned}
a_{m0} &= \frac{2m+1}{4\pi}\int\limits_C P_m(\cos\gamma)P_m(\cos\varphi)\,\mathrm{d}\sigma = P_m(\cos\varphi'),\\
a_{mp} &= 2\frac{(m-p)!}{(m+p)!}P_m^p(\cos\varphi')\cos(p\theta') \qquad (p=1,2,\cdots,m),\\
b_{mp} &= 2\frac{(m-p)!}{(m+p)!}P_m^p(\cos\varphi')\sin(p\theta') \qquad (p=1,2,\cdots,m).
\end{aligned} \tag{5.170}$$

Theorem 5.17.1 (Addition Theorem) *Given the above expressions of the series coefficients, the explicit expansion of* $P_m(\cos\gamma)$ *is*

$$\begin{aligned}
&P_m(\cos\gamma) = P_m(\cos\varphi')P_m(\cos\varphi)+\\
&+2\sum_{p=1}^{m}\frac{(m-p)!}{(m+p)!}P_m^p\left[\cos\varphi)P_m^p(\cos\varphi')(\cos(p\theta)\cos(p\theta')+\sin(p\theta)\sin(p\theta')\right],
\end{aligned}$$

which is equivalent, applying the trigonometric identity $\cos\left[p(\theta-\theta')\right] = \cos(p\theta)\cos(p\theta')+\sin(p\theta)\sin(p\theta')$, *to*

$$\begin{aligned}
P_m(\cos\gamma) &= P_m(\cos\varphi')P_m(\cos\varphi)+\\
&+2\sum_{p=1}^{m}\frac{(m-p)!}{(m+p)!}P_m^p(\cos\varphi)P_m^p(\cos\varphi')\cos[p(\theta-\theta')].
\end{aligned}$$

Note 17
Clearly, (θ, φ) and (θ', φ') are interchangeable in $P_m(\cos\gamma)$.

We can now calculate the generic $U_m(\theta', \varphi')$ in the expansion of the potential. Actually, it is

$$\begin{aligned} U_m(\theta', \varphi') &= G \int\int_C\int P_m(\cos\gamma) r^m \rho(\mathbf{r})\, \mathrm{d}^3\mathbf{r} \\ &= G\Bigg[P_m(\cos\varphi') \int_C P_m(\cos\varphi)\, r^m\, \mathrm{d}m + \\ &\quad + 2\sum_{p=1}^{m} \frac{(m-p)!}{(m+p)!} P_m^p(\cos\varphi') \int_C P_m^p(\cos\varphi) \cos[p(\theta-\theta')]\, r^m\, \mathrm{d}m\Bigg]. \end{aligned} \tag{5.171}$$

Let us consider, first, the first addend in square bracket in Eq. 5.171

$$\begin{aligned} P_m(\cos\varphi') \int_C P_m(\cos\varphi)\, r^m\, \mathrm{d}m &= M R^m P_m(\cos\varphi') \int_C P_m(\cos\varphi) \left(\frac{r}{R}\right)^m \frac{\mathrm{d}m}{M} \equiv \\ &\equiv -M R^m P_m(\cos\varphi') J_m, \end{aligned} \tag{5.172}$$

having defined the dimensionless quantity

$$J_m \equiv -\int_C P_m(\cos\varphi) \left(\frac{r}{R}\right)^m \frac{\mathrm{d}m}{M}, \tag{5.173}$$

which is, indeed, the opposite of the (dimensionless) mth moment of r weighted with $\rho(\mathbf{r}) P_m(\cos\varphi)$. Consequently, the corresponding generic mth contribution to the external potential is so

$$-\frac{GM}{r'} \left(\frac{R}{r'}\right)^m J_m P_m(\cos\varphi'). \tag{5.174}$$

Let us now consider the other contribution to U_m, that coming from the sum over p in the square brackets of Eq. 5.171, whose generic term is[19]

$$2\frac{(m-p)!}{(m+p)!}P_m^p(\cos\varphi')\int_C P_m^p(\cos\varphi)\cos[p(\theta-\theta')]r^m\,dm =$$

$$= 2MR^m\frac{(m-p)!}{(m+p)!}P_m^p(\cos\varphi')\Bigg[\cos(p\theta')\int_C P_m^p(\cos\varphi)\cos(p\theta)\left(\frac{r}{R}\right)^m\frac{\mathrm{d}m}{M}+$$

$$+\sin(p\theta')\int_C P_m^p(\cos\varphi)\sin(p\theta)\left(\frac{r}{R}\right)^m\frac{\mathrm{d}m}{M}\Bigg] =$$

$$= MR^m P_m^p(\cos\varphi')\left(\cos(p\theta')J_{mp}+\sin(p\theta')d\bar{J}_{mp}\right), \tag{5.175}$$

where

$$\begin{aligned} J_{mp} &= 2\frac{(m-p)!}{(m+p)!}\int_C P_m^p(\cos\varphi)\cos(p\theta)\left(\frac{r}{R}\right)^m\frac{\mathrm{d}m}{M},\\ \bar{J}_{mp} &= 2\frac{(m-p)!}{(m+p)!}\int_C P_m^p(\cos\varphi)\sin(p\theta)\left(\frac{r}{R}\right)^m\frac{\mathrm{d}m}{M}, \end{aligned} \tag{5.176}$$

are dimensionless quantities.

Finally, the generic term in the external potential series expansion of Eq. 5.160 is

$$U_m\frac{1}{r'^{\,m+1}1} = \frac{GM}{r'}\left(\frac{R}{r'}\right)^m \times$$

$$\times\left[-J_m P_m(\cos\varphi') + \sum_{p=1}^{m}\left(J_{mp}\cos(p\theta')+\bar{J}_{mp}\sin(p\theta')\right)P_m^p(\cos\varphi')\right]. \tag{5.177}$$

The potential results to be a series where a radial, power decaying, part is modulated by a (θ',φ') dependence through surface spherical harmonics with coefficients depending on the distribution of matter density.

As expected, if we consider only the terms with $m = 0, 1$, and choose the origin of the reference in the centre of mass, $O \equiv G$, the expansion reduces to

$$U(r') = \frac{GM}{r'}, \tag{5.178}$$

because $J_0 = -1,\ J_1 = J_{11} = \bar{J}_{11} = 0$.

[19] Using, again, $\cos[p(\theta-\theta')] = \cos(p\theta)\cos(p\theta') + \sin(p\theta)\sin(p\theta')$.

Exercise 5.10 Determine the expression by multipole series of the gravitational potential generated by a massive homogeneous ring Σ_R of mass $M = 2\pi\lambda_0 R$, where λ_0 is the (constant) linear mass density of the ring and R its radius.

Solution

$$(i) \quad \text{If} \quad r < R \quad \text{then} \quad U(x, y, z) = \sum_{m=0}^{\infty} U_{ext,m} r^m;$$

$$(ii) \quad \text{if} \quad r > R \quad \text{then} \quad U(x, y, z) = \sum_{m=0}^{\infty} U_{int,m} \frac{1}{r^{m+1}}.$$

Case *(i)*, $r < R$.
By virtue of the axial symmetry of the problem, once assumed $\varphi = \pi/2$ on the ring plane,

$$U_{ext,m} = G \int_{\Sigma_R} P_m(\cos\gamma) \frac{1}{r^{m+1}} \, \mathrm{d}m = G P_m(\cos\varphi) \int_{\Sigma_R} P_m(\cos\varphi) \frac{1}{r^{m+1}} \, \mathrm{d}m =$$
$$= \frac{GM}{R^{m+1}} P_m(\cos\varphi) \int_{\Sigma_R} P_m(\cos\varphi) \left(\frac{R}{r}\right)^{m+1} \frac{\mathrm{d}m}{M} = -\frac{GM}{R^{m+1}} P_m \tilde{J}_m,$$

where

$$\tilde{J}_m = -\int_{\Sigma_R} P_m(\cos\varphi) \left(\frac{R}{r}\right)^{m+1} \frac{\mathrm{d}m}{M} = -P_m(0) =$$
$$= \begin{cases} -1 & \text{if } m = 0, \\ 0 & \text{if } m \text{ is odd}, \\ -(-1)^{m/2} \dfrac{\prod_{k=1}^{m/2} (2k-1)}{2^{m/2}\,(m/2)!} & \text{if } m \text{ is even}. \end{cases} \tag{5.179}$$

It turns out that

$$U(x, y, z) = \frac{GM}{R} \sum_{m=0}^{\infty} \left[-\tilde{J}_m P_m(\cos\varphi) \left(\frac{r}{R}\right)^m \right] =$$

$$= 2\pi G\lambda_0 \left[1 + \sum_{\substack{m=2 \\ m \text{ even}}}^{\infty} (-1)^{m/2} \frac{\prod_{k=1}^{m/2}(2k-1)}{2^{m/2}\,(m/2)!} P_m(\cos\varphi) \left(\frac{r}{R}\right)^m \right] =$$

$$= 2\pi G\lambda_0 \Bigg[1 - \frac{1}{2} P_2(\cos\varphi) \left(\frac{r}{R}\right)^2 + \frac{1\cdot 3}{2^2\cdot 2} P_4(\cos\varphi) \left(\frac{r}{R}\right)^4 +$$

$$- \frac{1\cdot 3\cdot 5}{2^3\cdot 3!} P_6(\cos\varphi) \left(\frac{r}{R}\right)^6 + \cdots \Bigg].$$

Case *(ii)*, $r > R$

$$U(x, y, z) = \frac{GM}{r} \left[1 - \sum_{\substack{m=2 \\ m \text{ even}}}^{\infty} J_m P_m(\cos\varphi) \left(\frac{R}{r}\right)^m \right];$$

upon the choice of the origin of the reference system in the centre of mass ($J_1 = 0$). The zonal coefficients are

$$J_m = -\int_{\Sigma_R} P_m(\cos\varphi) \left(\frac{r}{R}\right)^m \frac{\mathrm{d}m}{M} = -P_m(0) =$$

$$= \begin{cases} -1 & \text{if } m = 0, \\ 0 & \text{if } m \text{ is odd}, \\ -(-1)^{m/2} \dfrac{\prod_{k=1}^{m/2}(2k-1)}{2^{m/2}\,(m/2)!} & \text{if } m \text{ is even}. \end{cases} \tag{5.180}$$

so that

$$U(x, y, z) = \frac{GM}{r}\left[1 - \frac{1}{2}P_2(\cos\varphi)\left(\frac{R}{r}\right)^2 + \frac{1\cdot 3}{2^2\cdot 2}P_4(\cos\varphi)\left(\frac{R}{r}\right)^4 - \cdots\right] =$$

$$= 2\pi G\lambda_0\left[\frac{R}{r} - \frac{1}{2}P_2(\cos\varphi)\left(\frac{R}{r}\right)^3 + \frac{1\cdot 3}{2^2\cdot 2}P_4(\cos\varphi)\left(\frac{R}{r}\right)^5 + \right.$$

$$\left. - \frac{1\cdot 3\cdot 5}{2^3\cdot 3!}P_6(\cos\varphi)\left(\frac{R}{r}\right)^6 + \cdots\right].$$

Note that both the above series are not necessarily alternating series, because this characteristics depends on the sign of $P_m(\cos\varphi)$. For example, on the equatorial plane ($\varphi = \pi/2$), $P_m(0)$ is, for m even, alternating in sign in such a way that the product $J_m P_m(0) < 0$ for every even m and so the series is not alternating. On the contrary, on positive and negative polar axis ($\varphi = 0$ and $\varphi = \pi$, respectively) all the Legendre polynomial of even m give $P_m(\pm 1) = 1$ and so the series is alternating.

Exercise 5.11 Evaluate

(*i*) the gravitational force per unit mass exerted by the homogeneous ring considered in the previous exercise in points at distances $r > R$ from the ring centre, and,

(*ii*) determine the deviation of the radial acceleration from the simple monopole term when using the first 3 nonzero terms in the series expansion.

Solution

(*i*) In axisymmetry, $U = U(r, \varphi)$ is independent of θ, and the corresponding force per unit mass is

$$\mathbf{F} = \nabla U = \frac{\partial U}{\partial r}\mathbf{e}_r + \frac{1}{r}\frac{\partial U}{\partial \varphi}\mathbf{e}_\varphi,$$

where

$$F_r = \frac{\partial U}{\partial r} = -\frac{GM}{r^2}\left[1 - \sum_{\substack{m=2\\ m\ \text{even}}}^{\infty} J_m P_m(\cos\varphi)\left(\frac{R}{r}\right)^m \right.$$

$$\left. - \sum_{\substack{m=2\\ m\ \text{even}}}^{\infty} J_m P_m(\cos\varphi) m\left(\frac{R}{r}\right)^m\right] = \qquad (5.181)$$

$$= -\frac{GM}{r^2}\left[1 - \sum_{\substack{m=2\\ m\ \text{even}}}^{\infty}(m+1)J_m P_m(\cos\varphi)\left(\frac{R}{r}\right)^m\right].$$

The summation over even values of $m \geq 2$ in the above equation corresponds to

$$m = 2l,\ l \geq 1 \Longrightarrow (m+1)J_m = -(2l+1)(-1)^l \frac{\prod_{k=1}^{l}(2k-1)}{2^l l!}$$

where the fraction numerator is the product of the first l odd positive integers. It turns out that the radial force expansion is a series, whose first terms are

$$F_r = -\frac{GM}{r^2}\left[1 - \frac{3}{2}P_2(\cos\varphi)\left(\frac{R}{r}\right)^2 + \frac{15}{8}P_4(\cos\varphi)\left(\frac{R}{r}\right)^4 - \right.$$
$$\left. + \frac{105}{48}P_6(\cos\varphi)\left(\frac{R}{r}\right)^6 + \cdots\right]. \tag{5.182}$$

The series 5.182 is not, in general, an alternating series because of the presence of the Legendre polynomials.

The φ component of the force is

$$F_\varphi = \frac{1}{r}\frac{\partial U}{\partial \varphi} = \frac{GM}{r^2}\sum_{\substack{m=2 \\ m \text{ even}}}^{\infty} J_m P'_m(\cos\varphi)\sin\varphi\left(\frac{R}{r}\right)^m,$$

where the prime symbol means derivative with respect to $t = \cos\varphi$.

Expressing some terms, the series above gives

$$F_\varphi = \frac{GM}{r^2}\left[\frac{1}{2}P'_2(\cos\varphi)\sin\varphi\left(\frac{R}{r}\right)^2 - \frac{1\cdot 3}{2^2\cdot 2}P'_4(\cos\varphi)\sin\varphi\left(\frac{R}{r}\right)^4 + \right.$$
$$\left. + \frac{1\cdot 3\cdot 5}{2^3\cdot 3!}\sin\varphi\left(\frac{R}{r}\right)^6 - \cdots\right] = -\frac{3}{8}\frac{GM}{r^2}\sin(2\varphi)\left[-\left(\frac{R}{r}\right)^2 + \right.$$
$$+ \frac{1}{4}\left(35\cos^2\varphi - 15\right)\left(\frac{R}{r}\right)^4$$
$$\left. - \frac{5}{96}\left(693\cos^4\varphi - 630\cos^3\varphi + 105\right)\left(\frac{R}{r}\right)^6 + \cdots\right].$$

(*ii*) The relative difference of the radial component of the force per unit mass to the monopole term $-GM/r^2$ is (see Eq. 5.182)

$$s_r = \frac{F_r - \left(-\frac{GM}{r^2}\right)}{\frac{GM}{r^2}} = \sum_{\substack{m=2 \\ m \text{ even}}}^{\infty} (m+1)J_m P_m(\cos\varphi)\left(\frac{R}{r}\right)^m.$$

Table 5.5 Deviations from the monopole acceleration (2nd column) at various distances from the ring centre on the z axis (1st column). Third column gives the error bound using 3 nonzero terms in the series expansion

R/r	s_r	$\lvert\epsilon_3\rvert$
1/2	0.292	0.00961
1/10	0.01482	2.46×10^{-6}
1/100	0.0001498	2.46×10^{-14}

As we said, in general at varying φ the terms in the above series are not necessarily alternating in sign, but for $\varphi = 0$ and $\varphi = \pi$ (polar axis) they are indeed alternating, because $P_m(1) = 1$ for every even m and J_m, as given by Eq.5.180, has alternating sign. Because the series is alternating and its general nonzero ($m = 2l, l = 1, 2, 3, \cdots$) term

$$a_{2l} = (2l+1)\frac{(2l-1)!!}{2^l l!}\left(\frac{R}{r}\right)^{2l}$$

is positive, decreasing, and tends to zero as m increases, we can use the Leibnitz criterion to evaluate the bound R_L to the error committed considering the first L terms in the series for s_r, to have

$$|\epsilon_L| \le a_{2(L+1)} = (2L+3)|J_{2(L+1)}|\left(\frac{R}{r}\right)^{2(L+1)}.$$

Given all this, it can be seen that taking as examples $R/r = 1/2$, $R/r = 1/10$ and $R/r = 1/100$ it results (using the $m = 2, 4, 6$ terms) a deviation from monopole as given in Table 5.5, where also the error bound using the Leibnitz criterion above is reported.

Exercise 5.12 Given a spherical massive shell of uniform surface mass density σ_0 and radius R, it is asked to show that the gravitational force vanishes inside the shell.

Solution The potential series expansion in points internal to the shell ($r < R$) is

$$U(x, y, z) = \frac{GM}{R}\sum_{m=0}^{\infty}\left[-\tilde{J}_m P_m(\cos\varphi)\left(\frac{r}{R}\right)^m\right],$$

where

$$\tilde{J}_m = -\int\!\!\int_{\Sigma_R} P_m(\cos\varphi)\left(\frac{R}{r}\right)^{m+1} \frac{\sigma_0 r^2 \sin\varphi \, \mathrm{d}\theta \mathrm{d}\varphi}{M} =$$

$$= \frac{\sigma_0 R^2 2\pi}{M} \int_{-1}^{1} P_m(t)\, dt = \begin{cases} 1, \text{ if } m = 0, \\ 0, \ \forall m \neq 0, \end{cases}$$

so that

$$U(x, y, z) = \frac{GM}{R} = \text{constant} \quad \Longrightarrow \quad \mathbf{F} = \nabla U = 0, \quad \forall r < R.$$

Note that this is an alternative demonstration of the 1st Newton's theorem.

Exercise 5.13 Show that any radial perturbation $\delta U(r)$ to a Keplerian potential leads to a precession of the motion of the unit mass particle. Find the precession angle in the case of a radial correction to the monopole term of the form $\delta U(r) = \alpha/r^2$.

Solution Given a spherical potential $U_0(r)$ a perturbation leads, in general, to

$$U(r, \theta, \varphi) = U_0(r) + \delta U(r, \theta, \varphi) = \frac{k}{r} + \delta U(r, \theta, \varphi).$$

where the last equality holds in the assumption of an unperturbed Keplerian potential, $U_0(r) \equiv k/r$, with $k > 0$. Of course, a purely radial perturbation is such that δU depends only on r.

First, note that the condition for the perturbation to be small is

$$\left|\frac{\delta U}{U_0}\right| \ll 1,$$

that defines the region of space where linear considerations are valid. For example, assuming $\delta U(r) = \alpha/r^\beta$ with α a non-zero constant and $\beta > 0$, the previous inequality leads to $r^{\beta-1} \gg |\alpha|/k$. If $\beta = 2$, the radial perturbation is small only in regions sufficiently out of the sphere of radius $R = |\alpha|/k$.

Let us evaluate the precession angle $\Delta\theta$, defined in Chap. 3, Sect. 3.7, as

$$\Delta\theta = 2L \int_{r_-}^{r_+} \frac{1}{r^2} \frac{\mathrm{d}r}{\sqrt{2\left(E + U - \frac{1}{2}\frac{L^2}{r^2}\right)}}.$$

To transform the integrand from an irrational to a rational function, a formal derivation and an indefinite integration over L is performed, to give

$$\Delta\theta = \frac{\partial}{\partial L}\int dL\, 2L\left[\int_{r_-}^{r_+}\frac{1}{r^2}\frac{\mathrm{d}r}{\sqrt{2\left(E+U-\frac{1}{2}\frac{L^2}{r^2}\right)}}\right]$$
$$= -2\frac{\partial}{\partial L}\int_{r_-}^{r_+}\sqrt{2\left(E+U-\frac{1}{2}\frac{L^2}{r^2}\right)}\,\mathrm{d}r.$$

Letting $f(U) \equiv \sqrt{2\left(E+U-\frac{L^2}{2r^2}\right)}$, its expansion up to 1st order in U around U_0 is[20]

$$f(U) = f(U_0) + f'(U_0)\delta U = \sqrt{2\left(E+\frac{k}{r}-\frac{1}{2}\frac{L^2}{r^2}\right)} + \frac{\delta U}{\sqrt{2\left(E+\frac{k}{r}-\frac{1}{2}\frac{L^2}{r^2}\right)}},$$

by which

$$\Delta\theta = \underbrace{-2\frac{\partial}{\partial L}\int_{r_-}^{r_+} f(U_0)\,\mathrm{d}r}_{Keplerian=2\pi}\ \underbrace{-2\frac{\partial}{\partial L}\int_{r_-}^{r_+}\frac{\delta U}{\sqrt{2\left(E+\frac{k}{r}-\frac{1}{2}\frac{L^2}{r^2}\right)}}\,\mathrm{d}r}_{\delta\theta}.$$

The first integral in the expression above is, indeed, simply the Keplerian θ variation along a full radial oscillation, i.e. 2π (the trajectory is fixed). The second integral above gives the variation $\delta\theta$ respect to the Keplerian value, so that $\Delta\theta = 2\pi + \delta\theta$. In Keplerian approximation and using the angular momentum conservation, it is (for $r_- \le r \le r_+$)

$$\frac{\mathrm{d}r}{\sqrt{2\left(E+\frac{k}{r}-\frac{1}{2}\frac{L^2}{r^2}\right)}} = \frac{r^2}{L}\mathrm{d}\theta.$$

so that

$$\delta\theta = -2\frac{\partial}{\partial L}\int_{\theta_-}^{\theta_+}\delta U\frac{r^2}{L}\,\mathrm{d}\theta.$$

[20] The prime, $'$, stands for derivative with respect to U_0.

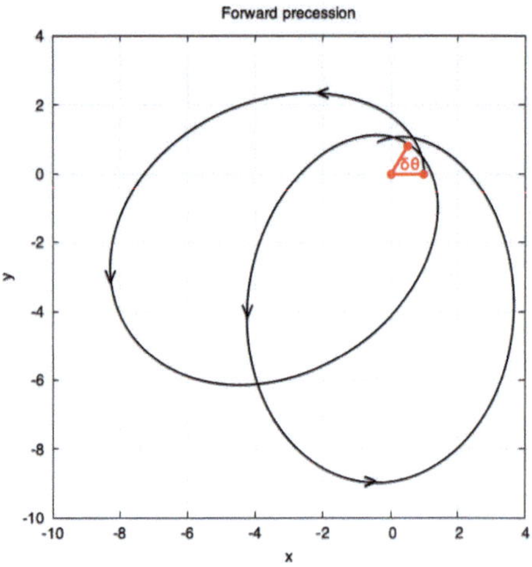

Fig. 5.10 Forward precession ($\alpha > 0$ in the perturbing potential $\delta U(r)$). At $t = 0$ the particle starts from pericentre on the x-axis ($x = 1$, $y = 0$). In red the $\delta\theta$ advancement after the 1st radial oscillation

An explicit computation of $\delta\theta$ would require the knowledge of both $\theta_- \equiv \theta(r_-)$ and $\theta_+ \equiv \theta(r_+)$, thing that implicitly requires the knowledge of the solution of the motion trajectory problem. The perturbation $\delta U(r) = \alpha/r^2$ leads to

$$\delta\theta = -2\alpha\,[\theta(r_+) - \theta(r_-)]\,\frac{\partial}{\partial L}\frac{1}{L} = \frac{2\alpha}{L^2}\,[\theta(r_+) - \theta(r_-)]\,.$$

Assuming for $\theta_\pm$ their Keplerian values, $\theta_- = 0$ and $\theta+ = \pi$, we finally have

$$\Delta\theta = 2\pi\left(1 + \frac{\alpha}{L^2}\right).$$

Clearly,

- If $\alpha > 0$ the precession is *forward* (counterclockwise rotation of the apsidal line[21] respect to the fixed reference frame), because $\Delta\theta > 2\pi$ (see Fig. 5.10).
- If $\alpha < 0$ the precession is *backward* (clockwise rotation of the apsidal line), because $\Delta\theta < 2\pi$.

The backward and forward character of the precession is easily interpreted. Actually, the radial force exerted by the perturbed potential on the unitary mass (radial acceleration) is

$$\mathbf{F}_r = \nabla U = -\left(\frac{k}{r^2} + \frac{2\alpha}{r^3}\right)\mathbf{e}_r = -\frac{k}{r^2}\left(1 + \frac{2\alpha}{kr}\right)\mathbf{e}_r,$$

so that the ratio between the actual perturbed force and the Keplerian is

[21] The apsidal line is the line connecting the pericentre to apocentre at every time.

$$\left|\frac{\mathbf{F}_r}{k/r^2}\right| = \left|1 + \frac{2\alpha}{k}\frac{1}{r}\right|,$$

and, consequently, the curvature of the test particle trajectory (which is proportional to the acceleration) is larger (smaller) than the local Keplerian trajectory when α is positive (negative), inducing a forward or backward precession in the two cases, respectively.

5.18 The Axisymmetric Earth Potential

Let $O \equiv G$ (centre of mass) be the origin of a Cartesian reference frame with the z-axis parallel to the Earth angular rotation velocity, and the equatorial xy-plane such that the x-axis points to the prime meridian. Moreover, we indicate by R the equatorial semiaxis.

According to geophysics data, the internal Earth mass density does not show a significant departure from axisymmetry, i.e. $\partial\rho/\partial\theta \simeq 0$, with our definition of r,θ,φ. Consequently, it turns out that (Eq. 5.176)

$$J_{mp} = \bar{J}_{mp} \simeq 0. \tag{5.183}$$

The above result is due to the fact that in axisymmetry (ρ depends only on r and φ), the θ dependence separates from r and φ into the integrals in Eq.5.176, giving

$$J_{mp} = 2\frac{(m-p)!}{(m+p)!}\frac{1}{M}\int_0^{\pi}\int_0^{r(\varphi)} P_m^p(\cos\varphi)\left(\frac{r}{R}\right)^m \rho(r,\varphi)\, r^2 \sin\varphi \,\mathrm{d}r\mathrm{d}\varphi \int_0^{2\pi}\cos(p\theta)\,\mathrm{d}\theta, \tag{5.184}$$

where (for every p)

$$\int_0^{2\pi}\cos(p\theta)\,\mathrm{d}\theta = 0. \tag{5.185}$$

The same holds for every $\bar{J}_{mp}$, which contains $\int_0^{2\pi}\sin(p\theta)\,\mathrm{d}\theta$, which is zero, too.

So, the general result for the external Earth potential in axisymmetry is

$$U(x', y', z') = \frac{GM}{r'}\left[1 - \sum_{m=2}^{\infty}\left(\frac{R}{r'}\right)^m J_m P_m(\cos\varphi')\right], \tag{5.186}$$

meaning that the potential generated by an axisymmetric matter distribution is expressed in terms of zonal harmonics, only.

Let us determine the quadrupole term, J_2, which is written as

$$
\begin{aligned}
J_2 &= \int_C P_2(\cos\varphi)\left(\frac{r}{R}\right)^2 \frac{\mathrm{d}m}{M} = \frac{1}{M}\int_C \left(\frac{1}{2}-\frac{3}{2}\cos^2\varphi\right)\left(\frac{r}{R}\right)^2 \mathrm{d}m = \\
&= \frac{1}{MR^2}\frac{1}{2}\left(\int_C r^2\,\mathrm{d}m - 3\int_C r^2\cos^2\varphi\,\mathrm{d}m\right) = \frac{1}{2}\frac{1}{MR^2}\left(I - 3\int_C z^2\,\mathrm{d}m\right),
\end{aligned}
\tag{5.187}
$$

where $I = \int_C r^2\,\mathrm{d}ms$ is the moment of inertia about the origin. If we denote by I_1, I_2, I_3 the axial moments of inertia with respect to the x-, y- and z-axes, respectively, which can be written as

$$
I_1 = \int_C (y^2+z^2)\,\mathrm{d}m, \quad I_2 = \int_C (x^2+z^2)\,\mathrm{d}m, \quad I_3 = \int_C (x^2+y^2)\,\mathrm{d}m, \tag{5.188}
$$

we have

$$
I = \frac{1}{2}(I_1 + I_2 + I_3), \tag{5.189}
$$

and

$$
\int_C z^2\,\mathrm{d}m = I - I_3, \tag{5.190}
$$

hence

$$
\begin{aligned}
J_2 &= \frac{1}{2}\frac{1}{MR^2}\left[I - 3(I - I_3)\right] = \frac{1}{2}\frac{1}{MR^2}\left[2I_3 - (I_1 + I_2)\right] = \\
&= \frac{1}{MR^2}\left(I_3 - \frac{I_1 + I_2}{2}\right) > 0.
\end{aligned}
\tag{5.191}
$$

The coefficient J_2 is positive because the z-axis is along the Earth minor axis (which is along its spin direction) so that $I_3 > I_2$ and $I_3 > I_1$.

From satellite measurements of M, R, I_1, I_2, I_3, the value $J_2 = 1.0827 \times 10^{-3}$ is obtained (see also Table 5.6).

5.19 Satellite Motion

The vector equation of motion for a point mass satellite in the axisymmetric Earth field, neglecting atmospheric drag, is

$$\ddot{\mathbf{r}} = \nabla U = \nabla \left\{ \frac{GM}{r} \left[1 - \sum_{m=2}^{\infty} \left(\frac{R}{r} \right)^m J_m P_m(\cos\varphi) \right] \right\}. \tag{5.192}$$

The above potential $U(\mathbf{r})$ can be thought of as a *perturbed* monopole potential

$$U(x, y, z) = \frac{GM}{r} + \mathcal{R}, \tag{5.193}$$

where the perturbing function $\mathcal{R}$ is

$$\mathcal{R}(r, \varphi) = -\frac{GM}{r} \sum_{m=2}^{\infty} \left(\frac{R}{r} \right)^m J_m P_m(\cos\varphi). \tag{5.194}$$

The above function actually constitutes a small perturbation over the monopole term, because (see Table 5.6) J_2 (which is by far the largest of the J_ms) is only $\sim 10^{-3}$.

Using the perturbing function, the full 1st order system of equations of motion for the satellite is

$$\begin{cases} \dot{x} = u \\ \dot{y} = v \\ \dot{z} = w \\ \dot{u} = -\dfrac{GM}{r^3} x + \dfrac{\partial \mathcal{R}}{\partial x} \\ \dot{v} = -\dfrac{GM}{r^3} y + \dfrac{\partial \mathcal{R}}{\partial y} \\ \dot{w} = -\dfrac{GM}{r^3} z + \dfrac{\partial \mathcal{R}}{\partial z}, \end{cases} \tag{5.195}$$

with initial conditions

$$\begin{cases} x(t_0) = x_0 \\ y(t_0) = y_0 \\ z(t_0) = z_0, \end{cases} \qquad \begin{cases} u(t_0) = \dot{x}_0 \\ v(t_0) = \dot{y}_0 \\ w(t_0) = \dot{z}_0. \end{cases} \tag{5.196}$$

Of course, in the limit $\mathcal{R} \to 0$ the satellite motion is Keplerian.

The magnitude of the relative perturbation over the monopole, $|\mathcal{R}/(GM/r)|$, is of order J_2, so that what it is expected is a *secular* (i.e. occurring on a time scale significantly longer than the Keplerian orbital period) departure of the actual orbit from the Keplerian, unperturbed, one.

The solution of the system 5.195 is modernly reachable at the highest precision with numerical methods. Anyway, it is useful a short treatment of the classical perturbation method behind this problem.

5.20 A Short Note on the Perturbation Theory: Osculating Elements

The unperturbed solution of the equation of motion is by definition that obtained for $\mathcal{R} = 0$. It depends on the six constants (x_0,y_0,z_0) and ($\dot{x}_0$,$\dot{y}_0$,$\dot{z}_0$). Usually, another set of six constants is used at their place, called *elliptic elements*, which are linked to the initial conditions. The adjective "elliptic" derives from that the unperturbed solution for negative orbital energy is a Keperian ellipse.

The elliptic elements are:

- Ω, the longitude of the ascending node. It is defined as the angle formed by the x axis of the plane of reference with the intersection line between the orbital plane and the reference plane (called *line of nodes*) in the direction of the *ascending* node which is the point where the satellite passes from negative to positive values of z coordinate;
- i, the *inclination* angle between the orbital plane and the reference plane ($0^\circ \leq i \leq 180^\circ$);
- a, the trajectory *semi-major* axis;
- ω, the argument of *perigee*;
- e, the trajectory *eccentricity*;
- τ, the time of passage of the satellite at its perigee.

The unperturbed solution is, so, fully identified by the above set of values, $\mathbf{r} = \mathbf{r}(t; \Omega, i, a, w, e, \tau)$. The solution of the perturbed problem consists in finding the set of elliptic element in the form of slowly varying time functions such that at any time t the values of the elliptic elements at that time define a Keplerian elliptic orbit that osculates the real satellite orbit.

5.21 Search for the Time-Dependent Elliptic Elements

Without entering in details, we sketch here the standard procedure to solve the perturbed satellite problem. We indicate by $(\alpha_1, \alpha_2, \alpha_3, \alpha_4, \alpha_5, \alpha_6) = (\Omega, i, a, \omega, e, \tau)$, the set of elliptic elements, supposed to depend, slowly, on time. The goal is the determination of such functions $\alpha_i(t)$, $i = 1, 2, \cdots, 6$, as satisfying Eq. 5.195 with i.c. given by Eq. 5.196, in the form of $\alpha_i(t) = \alpha_i^0 + s_i(t) + p_i(t) \quad i = 1, 2, \cdots, 6$, where $\alpha_i^0 = \alpha_i(t_0)$ is the osculating elliptic element evaluated at $t = t_0$. The function $s_i(t)$ is the *secular* part of the solution and depends on the osculating elliptic elements and on the J_m coefficients, other than, linearly, on t; $p_i(t)$ is the *periodic* part of the solution. Were $s_i(t)$ and

Table 5.6 Some values of J_m as obtained by the model EGM2008 (from Pavlis et al., "The development and evaluation of the Earth Gravitational Model 2008 (EGM2008)", Journal of Geophysical Research: Solid Earth, 117(B4), B04406

m	$J_m/10^{-6}$
2	1082.626
3	−2.5327
4	−1.6196
5	−0.2272
6	0.5406
7	0.1599
8	0.0589
9	−0.0223
10	−0.0718
11	−0.024
12	0.0106
13	0.0061
14	−0.0073

Table 5.7 Values of G, M, and R. GM and R are obtained by the model EGM2008

$GM = 3.986005 \times 10^6\ \text{cm}^3\ \text{sec}^{-2}$
$R = 6378.137$ km=6.378165×10^8 cm
$M = 5.97217 \times 10^{27}$ g
$G = 6.67430 \times 10^{-8}$ dyn cm^2 g^{-2}

$p_i(t)$ known, an observation of the satellite at an initial time t_0 would give the whole set of α_i^0 and so the actual unknowns would be the J_m quantities. Thus, many different observations of different satellites at different times could constitute a way to evaluate some of the J_m, as well as the product GM and R. and so of the monopole component J_0. In Tables 5.6 and 5.7 we report values obtained in this way.

A more refined study would take into account the dependence, even if small, of the Earth potential on θ, that causes the presence in the perturbation function $\mathcal{R}$ of some tesseral harmonics, which are hard to determine.

In conclusion, given the definition of *geoid* as the shape that the ocean surface would assume under the influence of Earth gravity alone (in the absence of other effects such as winds and tides), it corresponds to an equipotential surface of the effective potential

$$U_{eff}(x, y, z) = U(x, y, z) + \frac{1}{2}\bar{\omega}^2(x^2 + y^2) = c, \quad \text{in Cartesian coordinates,}$$
$$U_{eff}(r, \theta, \varphi) = U(r, \theta, \varphi) + \frac{1}{2}\bar{\omega}^2 r^2 \sin^2\varphi = c, \quad \text{in spherical polar coordinates} \tag{5.197}$$

where $\bar{\omega}$ is the angular rotation velocity of the Earth and c is a constant.

Available data on the Earth potential from study of satellite motions show that the geoid deviates from the reference ellipsoid of revolution around the polar axis for no more than ±100 metres. Note that Newton was the first to evaluate the value of the relative deviation of the Earth ellipsoid from sphericity obtaining the value $s = 1/230$ which is quite similar to modern determinations, $\sim\frac{1}{298.25}$.

5.22 Further Readings

This chapter presented fundamentals of potential theory and of harmonic functions theory, with application to the specific Newtonian gravitational fields. Although self-consistent and easy to understand without additional readings, the reader who wants to know more about potential theory, harmonic functions and the, related, field of partial differential equations, would find useful the references [3–6, 8].

The final part of the chapter gives a short spot of classical celestial mechanics theory of perturbation, as well as some geodetical information on the Earth potential. The reader who wish to go deeper in the topics of celestial mechanics may take advantage by both the classical text [7] and the more modern one [2]. For a deeper approach to the Earth gravity field see [1].

References

1. Caputo, M.: The Gravity Field of the Earth from Classical and Modern Methods. International Geophysics Series, vol. 10. Academic Press, New York (1967)
2. Fitzpatrick, R.: An Introduction to Celestial Mechanics. Cambridge University Press, Cambridge (2012)
3. Gilbart, D., Trudinger, N.S.: Elliptic Equations of Second Order. Springer, New York (1977)
4. Kellogg, O.D.: Foundations of Potential Theory. Dover, New York (1953)
5. Mc Robert, T.M.: Spherical Harmonics. Pergamon Press, Oxford (1967)
6. Mikhailov, V.P.: Partial Differential Equations. MIR Publishers, Moscow (1978)
7. Moulton, F. R.: An Introduction to Celestial Mechanics, 2nd rev. edn. Dover, New York (1970)
8. Vladimirov, V.S.: Equations of Mathematical Physics. MIR Publishers, Moscow (1984)

Chapter 6
Unsolved Problems in Newtonian Theory

In spite of its great achievements and its capacity to explain phenomena in the solar system environment, the Newtonian theory of gravitation shows some limits, especially in some regimes (strong fields). In this chapter we review some of these issues.

6.1 Successes and Limits of Newtonian Gravity Theory

The success of the classical Newtonian gravity theory is well known. This success was essentially based on the excellent match of the observed motion of bodies in the solar system with what was expected on the basis of the Newton's 2nd law of dynamics in the case of a force expressed as a $1/r^2$ interaction. As we said, Kepler's laws are fully accounted for by the solution of the classical Newtonian 2-body problem.

Another great achievement was the prediction, made by Edmund Halley (1646–1742), of the periodicity of about 76 yr of the comet that, because of his successful prediction of its passage in 1758, was named after him. Notably, Halley was one fundamental mentor of Newton because he was able to convince Newton to write the three-volume treatise *Philosophiae Naturalis Principia Mathematica* where, in the 3rd volume, he expressed his inverse square law of gravitation. The publication expenses of the book were paid by Halley.

Another relevant figure in supporting Newtonian theory was Alexis Clairaut (1713–1765) who made the calculations to predict with precision the passage of the above mentioned comet and confirmed the Newton prediction for the shape of the Earth as an ellipsoid. Moreover, Clairaut was the first to face the so-called 3-body problem also with the aim of explaining the motion of the Moon as subjected to the combined force of the Sun and the Earth. Thanks to Leibnizian calculus, Clairaut was able to schematize the 3-body problem as a system of four differential

R. A. Capuzzo Dolcetta, *Classical Newtonian Gravity and Potential Theory*,
UNITEXT for Physics, https://doi.org/10.1007/978-3-032-13036-5_6

equations which, anyway, allowed just an approximate solution of the problem. In particular, a difficult and still unsolved problem was the explanation of the apsidal precession of the Moon orbit around the Earth, which is two times larger than predicted by Newton himself on the basis of his inverse square distance theory. Clairaut worked hard on the topic and, initially, reached the conclusion that the explanation of the observed 8.85 yr period of the Moon orbit apsidal line comes from the addition to the $1/r^2$ Newton force of an extra force term proportional to $1/r^3$. This finding was criticized because it went against the robustly based $1/r^2$ attraction law. This, together with a harsh debate with Leonard Euler (1707–1783), moved Clairaut towards a further deepening of the problem that, eventually, led to the solution of the problem in the full $1/r^2$ force framework. This, of course, was another important element of confirmation of the validity of the Newtonian gravity theory. It was not the last, because of the striking discovery of a planet on the basis of the Newtonian calculation predictions made almost at the same time by Urbain Jean Joseph Le Verrier (1811–1877) and John Couch Adams (1819–1892). They indeed suggested that the observed anomalies in the motion of Uranus could have been explained best by assuming the existence of a perturbing Newtonian force provided by a still unobserved planet which was actually observed by German astronomers in 1846 and that was named Neptune.

After these enormous achievements, Newtonian gravitation theory fell into an unsolvable problem: that of the unexpectedly large movement of the position of Mercury perihelion.

Around 1843 astronomical observations, mainly conducted by Le Verrier in Paris, were good enough to determine a movement of the Mercury perihelion position with respect to the "fixed" stars of 574″ per tropical century. Given that in a spherical gravitational field (which is supposedly that of the spherical Sun) the orbital pericentre of Mercury should be fixed in space and reached periodically in time, the explanation of the movement must be in the combined role of the other planets on Mercury. Le Verrier made accurate computations of the perturbing contribution of the planet closest to Mercury (Venus) and of the most massive (Jupiter) to reach, in 1859, the conclusion that the expected movement should be less than the actually observed one (in his report in the *Annales de l'Observatoire* he gave a discrepancy of 38" per century). Pushed by previous success in the interpretation of the anomalous orbit of Uranus as caused by the existence of the perturbing planet Neptune, Le Verrier suggested, also in this case, the existence of some still not observed massive object or of a group of asteroids on orbits internal to that of Mercury. He also suggested Vulcano as the name of the hypothetical perturbing planet. He reached this conclusion after discarding the alternative possibility that the mass of Venus is 10% larger than the one estimated.

If one wants to save the $1/r^2$ Newtonian force, another possibility to explain the discrepancy is the assumption of *aspherical* mass distribution within the Sun. This would essentially correspond to a quadrupole term J_2 in the expansion of the Sun potential significantly larger than the actually measured one, which can be excluded.[1]

The final interpretation, before the advent of Einstein's General Relativity theory, was the assumption of an interaction force between an ith and a jth body different from Newton's in the form $F_{ij} \propto r^{-(2+\epsilon)}$, with ϵ small and positive. Asaph Hall (1829–1907), around the end of the XIXth century, found that $\epsilon = 0.00000016$ suffices to explain the Mercury precession. Later, Simon Newcomb (1835–1909) gave another, supposedly more precise, value of $\epsilon = 0.0000001574$. But this explanation fell because it was incompatible with the observations of Lunar perigee advancement. Apart from some ambiguity in the proportionality constant for a modified version of the inverse square law expression which can make the modified force between bodies be larger or smaller than the standard $1/r^2$ dependence, it was clear that some variation of the actual gravitation with respect to the standard and universally accepted Newtonian expression would be needed. And, quite clearly, such variation should have been such as to represent better the role of gravity near massive objects—strong field regime—and, at the same time, to reconcile with Newtonian force when the field is weaker. As a matter of fact this variation was provided by the advent in 1915 of the General Relativity Theory (GRT) by Albert Einstein, as we now go to explain.

The $1/r^2$ Newtonian interaction leads to two mathematical (and conceptual) problems, the so-called *ultraviolet* (*UV*) and *infrared* (*IR*) divergence. The first one corresponds to the mutual force between two massive points becoming arbitrarily large at vanishing distance. This is not a real bug of Newtonian gravitation because, clearly, the closest approach is when the two bodies touch each other: the point mass approximation is just an unphysical approximation.

The *IR* divergence, on the other side, leads to the so-called Bentley's paradox: if gravity interaction is unscreened and unlimited in distance, the universe would be unstable. In a finite universe this instability would result in a single point collapse. This, according to Newton, leads to accepting an infinite universe. In an infinite universe every star is in equilibrium because equally pulled in every direction. The equilibrium is acknowledged by Newton to be unstable, because any change in the mutual distance would break it. Newton explained that this instability does not actually occur because "a continual miracle is needed to prevent the Sun and the fixed stars from rushing together through gravity". Anyway, the necessity of this miracle is avoided by the fact that stars in the sky are not fixed, as it was thought at those times, but actually move.

Another paradox to face in the allegedly infinite universe is the Olbers' paradox. This paradox consists in that in a universe infinite and uniformly populated by stars the night should be bright because in every direction an observer looks it always

[1] The amount of precession for asphericity of the Sun is almost linear in J_2 so that the present estimate of J_2 is only able to justify 0.07 % of what needed.

will encounter the surface of a star along its line of sight. This is because the $\propto r^2$ increase of the number of stars exactly compensates the $\propto 1/r^2$ decay of the radiation intensity coming from the emitting stars. Since the gravitational force also decreases the same way, the paradox applies also to gravity: an infinite universe seems incompatible with the finite gravitational force acting in every known point of space.

But this, also, was not an intrinsic limitation of Newtonian theory, because the supposed instability would arise just in the hypothesis, commonly accepted at those times, that stars are fixed in space. In a dynamical universe it is the motion itself of stars that gives the overall stability.

One less technical but equally important limitation of Newtonian gravitation is that it is a time independent field theory. This means that the gravitational force exerted on a massive body located in a generic point in space is simply the sum of the pairwise contributions from all the other surrounding objects, that depend only on mutual distance. This corresponds to an *instantaneous* propagation of the force: whatever the distance between the given body and the others, when they move they immediately exert and feel the mutual force.

6.1.1 General Relativity Solution of the Mercury Perihelion Advance

The puzzle of the missing $43''$ angular precession of Mercury perihelion found a full explanation in Einstein's 1915 theory of General Relativity. It is out of the purposes of this book to expand concepts of GRT; we just limit to say that the essential difference between the Newtonian theory of gravitation and the Einsteinian's is that the fundaments of Newtonian theory—like the concepts of *absolute* time and *absolute* space, as well as the instantaneous propagation of gravity interaction—were criticized and abandoned by Einstein. He proposed, instead, a gravitational theory where space and time are not absolute and invariant but, rather, mutually linked each other to form a 4 dimensional *space-time*, where gravity is interpreted as a deformation of such space-time induced by the local presence of mass/energy. Einstein claimed in one of his 1905 works that mass and energy are simply "two different expressions of the same thing", linked by the fundamental $E = mc^2$ relation that he introduced in his article of presentation of the Special Theory of Relativity. His definition of c (speed of light in vacuum) as limiting speed of any movement or interaction means that his theory of gravitation excludes instantaneous gravity interaction.

In his 1915 paper Einstein presented the equations of GRT which consist of a very complicated and non-linear system of partial differential equations whose unknowns are the parameters defining the geometry of the space-time where the gravitational "charges" are located. Once found the solutions of the field equations, it is seen that the motion of objects in such a curved space-time occurs along the *geodesic* lines.[2] Although Relativity theory provides results more accurate than Newtonian theory, the differences are appreciable only when dealing with objects moving at high speed and/or in strong enough gravitational fields where, indeed, the corrections to Newtonian results may rise to a significant level. In the cases of velocities much lower than c and in a weak gravitational field—planets in the solar system with the notable exception of Mercury, interplanetary probes, stars in galaxies whenever not too close to possible galactic central supermassive black holes, etc.—a Newtonian approach is precise enough and benefits from the huge advantage of providing a much simpler mathematical tool than GRT. On the other side, when seeking for the enormous precision needed to point locations on the Earth via GPS satellite constellations, or when dealing with dynamical phenomena around black holes, a full GRT treatment is needed. To give an idea of complications intrinsic to GRT treatment, it suffices to remind that while in Newtonian theory the 2-body problem is explicitly solved, in GRT it requires an enormous computational effort.

Going back to Mercury perihelion advancement, it is possible to show that the angular shift, $\delta\theta$, of the periapsis line over one full radial orbital oscillation (from pericentre to apocentre and back), in a space-time defined by a massive point located in the origin, $r = 0$, and after letting $u = 1/r$, is given by

$$\delta\theta \equiv \Delta\theta - 2\pi = \frac{1}{\sqrt{r_S(u_3 - u_1)}} \int_0^1 \frac{dt}{\sqrt{(1-t^2)(1-k^2t^2)}} - 2\pi, \qquad (6.1)$$

where $\Delta\theta$ has been defined in Eq. 3.128, and $k^2 = (u_2 - u_1)/(u_3 - u_1) \geq 0$, with u_1, u_2, u_3 solutions of the cubic equation

$$r_S u^3 - u^2 + \frac{r_S}{a^2} + \frac{1}{b^2} - \frac{1}{a^2} = 0. \qquad (6.2)$$

In the equation above $r_S = 2Gm_\bullet/c^2$ is the so-called Schwarzschild's radius of the point mass $m_\bullet$, and $a = L/c$, $b = cL/E$ are constants dependent on the orbital angular momentum, L, and energy, E, per unit mass.

The integral in Eq. 6.1 is called a *complete elliptic integral of the first kind*, usually referred to as $K(k)$. Once determined the values of r_S, u_1, u_2, u_3 (the latter three in a numerical way), k is determined. Anyway, the function $K(k)$ must be evaluated numerically, which is relatively easy nowadays at any level of precision required, but surely not at Einstein's time. For this reason in his 1915 paper he approached the problem in an approximate way. He kept just the first term in a

[2] A geodesic line is the line connecting two points through the shortest path. In a flat space-time the geodesic is simply the straight line segment connecting the two points.

expansion of $\delta\theta$ as series of the argument $1/c^2$, that is

$$\delta\theta = \frac{6\pi G}{c^2}\frac{m_\bullet}{a(1-e^2)} = 0.1''/\text{period}, \tag{6.3}$$

assuming for the Sun mass $m_\bullet = M_\odot = 1.989 \times 10^{33}$ g, and $e = 0.2056$, $a = 5.791 \times 10^{12}$ cm for the eccentricity and semi-major axis of the osculating elliptical orbit of Mercury.

Being the Mercury radial period $T_r = 0.24$ yr, the expression above leads indeed to $\approx 43''$ per century. The following term in the series expansion of $\delta\theta$ would be proportional to $1/c^4$, being

$$\frac{3\pi G^2}{2c^4}\frac{m_\bullet^2(30+5e^2)}{a^2(1-e^2)^2}, \tag{6.4}$$

which leads to the very small ratio $\simeq 3.58 \times 10^{-10}$ between this 2nd order term and Einstein's (Eq. 6.3) one.

6.1.2 *Other Failures and Successes of Newtonian and Relativity Theories*

While the failure of Newton's theory regarding Mercury perihelion advancement was noted around mid XIX century and solved by GRT in 1915, another crucial support to GRT came from the phenomenon of light *deflection*. This was more a great success of GRT rather than an indirect failure of Newton's theory, as we will now explain.

Newton studied optics and the characteristics of light and in his three book series *Opticks: or, A treatise of the Reflexions, Refractions, Inflexions and Colours of light* (1704) proposed a corpuscular theory of light. He said that most of the experimental results concerning light could be explained by assuming that light is nothing else than a bunch of small particles propagating radially from a source. In his view these particles travel in space along straight lines without being perturbed by the environment. This is somewhat contradictory with his gravitation theory because the particles of light, having a mass (even if very small), should be accelerated by massive objects along their path so that these paths should be, under specific circumstances, curved. He likely thought that the mass of the light particle is so small that the gravitational force it experiences is negligible.

At those times nobody was aware of the phenomenon of light deflection. It was Johann Georg von Soldner (1776–1833) who applied Newton's theory to evaluate light deflection as possibly caused by the Sun on light particles passing by its surface and moving at speed c. He obtained a deflection of just $0.87''$. It is interesting to note that this value is almost the same obtained in 1911 by Einstein by application of his special theory of relativity, i.e. before developing GRT. As I said above, GRT is a geometric theory of gravitation that says that space and time are sort of deformable textures which deviate from flat to curve whenever mass (or energy) acts on it. Due to that, light, like everything else, travels along geodesics in the space-time, and so, when a mass is interposed between the source of light (e.g. a star) and an observer, the path of the light would be no more rectilinear but would follow a more or less pronounced curved trajectory. To tell the truth this does not add a qualitative enormous change with respect to Newton's theory because, as we said, Newton himself proposed that light is material and so would feel gravitational deflection, whose amount was estimated by von Soldner. As a matter of fact, the difference between Newtonian and Einsteinian result was quantitative. In a paper of 1915 Einstein evaluated in $1.75''$ the deflection caused by the Sun on the passing-by light. This result stimulated the attention of scientists.

Consequently, in 1919, two scientific expeditions conducted by Arthur Eddington (1882–1944) in the Principe island, west of Gabon, and by Andrew Claude de la Cherois Crommelin (1865–1939) at Sobral (north of Brazil) were able to photograph stars surrounding the Sun during a total solar eclipse. They measured the star position dislocation at the eclipse in a period when the Sun was not interposed between them and the observer and found a shift of $1.75''$, in striking agreement with Eisntein's GRT prediction. and about twice the Newtonian value.

It is worth citing another important success of GRT which is partly linked to the phenomenon of light deflection. It is the so-called *radar echo delay*, also called Shapiro effect, after the name of Irwin I. Shapiro (1929) who proposed it in 1964 as a further validation of GRT. Like light deflection, this delay can be interpreted in classical mechanics but it is GRT that explains with high precision the time excess of the instant of return of the radar signal sent towards a planet at the time of its passage near (in projection) to the Sun.

First precise measurements of the Shapiro effect have been done in the years 60 of the XXth century by sending radar signals to Venus when it was almost in conjunction with the Sun, and measuring the difference with respect to the standard expected delay $\Delta t = 2d/c$, where d is the distance to Venus. The Shapiro effect has practical relevance for space travel because even a tiny deviation from the classical delay may be significant for space probe control. A comparison of the key features of Newtonian and General Relativity theories is summarized in Table 6.1.

Table 6.1 Scheme of some characteristics of the two theories of gravitation (Abbreviations: Class. = Classical, Quant. = Quantum, Prop. Speed = Propagation Speed)

Theory	Class./Quant.	Range	Prop. Speed	Complication
Newton	Class.	Infinite	Infinite	Low
GRT	Class.	Infinite	Finite	High

6.2 Further Readings

This short chapter presented some discussion of limits of Newton's theory of gravitation and how they were overcome in the frame of special and general theories of relativity. A simple review on this topic is found in [5]. For a deeper insight to the Einstein's theory of gravitation look at [2, 3]. The justification of the missing quantity in the Mercury perihelion advancement is found in the classical [1] paper. The Shapiro effect is described in [4].

References

1. Einstein, A.: Erklärung der Perihelbewegung des Merkur aus der allgemeinen Relativitätstheorie, pp. 831–839. Sitzungsberichte der Königlich Preußischen Akademie der Wissenschaften, Berlin (1915)
2. Misner, C.W., Thorne, K.S., Wheeler, J.A.: Gravitation. W. H. Freeman and Company, New York (1973)
3. Poisson, E., Will, C.M.: Gravity: Newtonian, Post-Newtonian, Relativistic. Cambridge University Press, Cambridge (2014)
4. Shapiro, I.I.: Fourth Test of General Relativity. Phys. Rev. Lett. **13**(26), 789–791 (1964)
5. Unzicker, A.: Why do we still believe in Newton's Law? Facts, myths and methods in Gravitational Physics. arXiv:0702.0009 [gr-qc] (2007)

Appendix A
Two- and N-Body Dimensionless Motion Equations

For practical computational convenience, it is useful to transform the equations of motion of a classical, Newtonian, N-body system into a non-dimensional form, so that the solution can be partially scaled for various sets of initial parameters.

Let us use the simple 2-body case as paradigm. The 2nd order vectorial equation of the relative motion for two objects of masses m_1 and m_2 is

$$\mu \ddot{\mathbf{r}} = -G \frac{m_1 m_2}{r^3} \mathbf{r}, \tag{A.1}$$

where $\mu = m_1 m_2/(m_1 + m_2)$ is the reduced mass and $\mathbf{r} \equiv \mathbf{r}_2 - \mathbf{r}_1$ (or, equivalently $\mathbf{r} \equiv \mathbf{r}_1 - \mathbf{r}_2$) gives the relative position of the two objects, identified by $\mathbf{r}_1$ and $\mathbf{r}_2$ in an inertial reference frame.

We can now define new, dimensionless, variables, $\mathbf{r}' = \mathbf{r}/L$, $m_i' = m_i/M$, $t' = t/T$, where $M = m_1 + m_2$, $T = L^{3/2}/\sqrt{GM}$. If $E < 0$, $T/(2\pi)$ is the orbital period (note also that the adoption of M, L and T as units of measure implies $G = 1$). This choice of units transforms Eq. A.1 into the dimensionless form

$$\frac{\mathrm{d}^2 \mathbf{r}'}{\mathrm{d}t'^2} = -\frac{\mathbf{r}'}{r'^3}. \tag{A.2}$$

Such procedure to get dimensionless equations can be followed also for the generic N-body case, where the equations of motion are (see 2.1.3.1), for every $1 \leq i \leq N$,

$$\begin{cases} \ddot{\mathbf{r}}_i = G \sum\limits_{\substack{j=1 \\ j \neq i}}^{N} m_j \dfrac{\mathbf{r}_j - \mathbf{r}_i}{r_{ij}^3}, \\ \mathbf{r}_i(0) = \mathbf{r}_{i0}, \\ \dot{\mathbf{r}}_i(0) = \dot{\mathbf{r}}_{i0}, \end{cases} \tag{A.3}$$

R. A. Capuzzo Dolcetta, *Classical Newtonian Gravity and Potential Theory*,
UNITEXT for Physics, https://doi.org/10.1007/978-3-032-13036-5

where $r_{ij} = \|\mathbf{r}_i - \mathbf{r}_j\|$ is the Euclidean distance between particle i and j.

The above system of $3N$ second-order scalar differential equations is reduced at first order by letting $\mathbf{v}_i = \dot{\mathbf{r}}_i$ $(i = 1, 2, \ldots, N)$ to have

$$\begin{cases} \dot{\mathbf{r}}_i = \mathbf{v}_i, \\ \dot{\mathbf{v}}_i = G \sum\limits_{\substack{j=1 \\ j \neq i}}^{N} m_j \dfrac{\mathbf{r}_j - \mathbf{r}_i}{r_{ij}^3}, \\ \mathbf{r}_i(0) = \mathbf{r}_{i0}, \\ \mathbf{v}_i(0) = \dot{\mathbf{r}}_{i0}. \end{cases} \tag{A.4}$$

Once formulated as a system of $6N$ first-order differential equations, its numerical solution can benefit from a plethora of existing algorithms.

As a matter of fact, only a very careful numerical treatment allows one to obtain a reliable solution, overcoming the *UV* and *IR* divergence of the interaction force and the strong non-linearity of the system of equations. I limit here to say that a straightforward way to overcome the problem of the divergent acceleration when two of the N bodies approach at vanishing distance is both to regularize the two-body interaction as described in Appendix B and to *soften* the purely Newtonian interaction potential. This means that the $U_{ij} \propto 1/r_{ij}$ interaction potential is replaced by a "softened", non-singular, potential. One way to soften the interaction potential is (for instance) via the substitution of the purely Newtonian pair interaction potential, $U_{ij} \propto 1/r_{ij}$ with the Plummer one described in Sect. 3.4.1 of this book, which has the form $U_{ij} \propto 1/\sqrt{r^2 + \epsilon^2}$, where ϵ (the constant b of the Plummer sphere) is the *softening* parameter (the length scale of potential smoothing). To have a simultaneous preservation of a proper global Newtonian behaviour of the N-body system and elimination of the small scale divergence, a proper choice of ϵ must be done. It is easily seen that a proper choice is a fraction of the average *closest neighbour* distance, $\langle d_{cn} \rangle \sim R/\sqrt[3]{N}$, where R is the physical size of the N-body system.

Regarding the topic of N-body dynamics, we limit ourselves to citing the book [1], which comprehensively covers most aspects of the challenging topic of gravitational N-body simulations.

Appendix B
Two-Body Regularization

The Newtonian interaction potential between two point mass particles, $U_{ij} \propto 1/r_{ij}$, diverges for two particles approaching zero distance (called *UV* divergence). This causes a relevant problem in the numerical integration of the evolution of an N-body system because, in order to follow with sufficient accuracy the rare close encounters, the time step must decrease by orders of magnitude (to the order of the fly-by time of the encounter, $t_{fb} \sim 2d/v_r$, where d is the closest approach distance and v_r is the two-body relative speed). As a consequence, the overall integration becomes computationally overwhelming, unless particular treatments are introduced (for example, methods apt to separate the i, j binary motion from the rest of the system, whose motion is essentially *frozen* over t_{fb}, to be resynchronized afterwards).

Another interesting, and more correct, way to deal with the UV divergence of U_{ij} is via its *regularization*, a procedure first introduced by the mathematician Levi-Civita and, later, faced and refined in many different ways by other researchers, among whom Kustaanheimo, Stiefel, etc.

Let us consider two bodies of masses m_i and m_j and position vectors $\mathbf{r}_i$ and $\mathbf{r}_j$ in a given reference frame. As explained in Appendix A, the equation of their relative motion can be written in non-dimensional form[1]

$$\ddot{\mathbf{r}} = -\frac{\mathbf{r}}{r^3}. \tag{B.1}$$

As said above, the force is singular in $r = 0$. By introducing the differential space-time transformation $\mathrm{d}t = r^n \mathrm{d}\tau$ (with a given but arbitrary n) which is equivalently written as $\mathrm{d}\tau = r^{-n}\mathrm{d}t$, we obtain

[1] The choices $G = 1$ and $m_i + m_j = 1$ imply that the unit of time is such that the orbital period of the bound 2-body motion is equal to $4\pi^2$.

R. A. Capuzzo Dolcetta, *Classical Newtonian Gravity and Potential Theory*,
UNITEXT for Physics, https://doi.org/10.1007/978-3-032-13036-5

$$\dot{\mathbf{r}} = \frac{\mathrm{d}\mathbf{r}}{\mathrm{d}\tau}\frac{\mathrm{d}\tau}{\mathrm{d}t} = \frac{\mathrm{d}\mathbf{r}}{\mathrm{d}\tau}r^{-n}, \tag{B.2}$$

so that the final, 2nd order, differential (with respect to τ) equation of motion is

$$\frac{\mathrm{d}^2\mathbf{r}}{\mathrm{d}\tau^2} - nr^{-1}\frac{\mathrm{d}r}{\mathrm{d}\tau}\frac{\mathrm{d}\mathbf{r}}{\mathrm{d}\tau} = -r^{2n-3}\mathbf{r}. \tag{B.3}$$

Choosing $n = 1$ and indicating with a prime, $'$, differentiation with respect to τ, the above equation takes the form

$$\mathbf{r}'' - \frac{1}{r}r'\mathbf{r}' = -\frac{\mathbf{r}}{r}. \tag{B.4}$$

For didactic convenience, we limit here to develop the one-dimensional case, letting thus $\mathbf{r} = x\,\mathbf{i}$, so that the above equation reduces to a scalar differential equation

$$x'' - \frac{x'^2}{x} = -1. \tag{B.5}$$

By means of the energy integral, $E = \frac{1}{2}\dot{x}^2 - \frac{1}{x} = \frac{1}{2}\left(\frac{x'}{x}\right)^2 - \frac{1}{x}$, we can substitute $\frac{x'^2}{x} = 2(Ex + 1)$ in Eq. B.5 obtaining

$$x'' - 2Ex = 1, \tag{B.6}$$

which is non-singular. A further change of variable, $u = 2Ex + 1$, transforms Eq. B.6 into a classic harmonic oscillator equation in the unknown u

$$u'' - 2Eu = 0, \tag{B.7}$$

whose solution in terms of sine and cosine functions is well known. The sought function $x(t)$ is finally obtained by the formula $x(\tau) = [u(\tau) - 1]/(2E)$, with the $\tau(t)$ relation coming from the inversion of the integral relation $t(\tau) = \int_0^\tau x(\tau)\mathrm{d}\tau$.

Reference

1. Aarseth, S.J.: Gravitational N-Body Simulations, Tools and Algorithms. Cambridge Monographs on Mathematical Physics. Cambridge University Press, Cambridge (2001)

Index

R. A. Capuzzo Dolcetta, *Classical Newtonian Gravity and Potential Theory*,
UNITEXT for Physics, https://doi.org/10.1007/978-3-032-13036-5

MIX
Papier aus verantwortungsvollen Quellen
Paper from responsible sources
FSC® C105338

If you have any concerns about our products,
you can contact us on
ProductSafety@springernature.com

In case Publisher is established outside the EU,
the EU authorized representative is:
Springer Nature Customer Service Center GmbH
Europaplatz 3, 69115 Heidelberg, Germany

Printed by Libri Plureos GmbH
in Hamburg, Germany